Coherent and Nonlinear Lightwave Communications

For a complete listing of the *Artech House Optoelectronics Library*, turn to the back of this book.

Coherent and Nonlinear Lightwave Communications

Milorad Cvijetic

Artech House
Boston • London

Library of Congress Cataloging-in-Publication Data
Cvijetic, Milorad.
   Coherent and nonlinear lightwave communications/Milorad Cvijetic.
      p.   cm.
   Includes bibliographical references and index.
   ISBN 0-89006-590-X (alk. paper)
1. Optical communications. 2. Laser communication systems. 3. Nonlinear optics. I. Title.
TK5103.59.C76   1996
621.382'7—dc20   95-49986
                                                                        CIP

British Library Cataloguing in Publication Data
Coherent and nonlinear lightwave communications
1. Laser communication systems 2. Optical communications 3. Integrated optics
I. Cvijetic, Milorad
621.3'827

ISBN 0-89006-590-X

© 1996 ARTECH HOUSE, INC.
685 Canton Street
Norwood, MA 02062

International Standard Book Number: 0-89006-590-X
Library of Congress Catalog Card Number: 95-49986

10 9 8 7 6 5 4 3 2 1

Contents

Preface

Coherent and nonlinear optical fiber transmission systems have been attracting great attention because they can provide the most efficient use of the enormous frequency bandwidth of single-mode optical fibers. Coherent and nonlinear lightwave communications are the most challenging research area in modern telecommunications.

This book is an attempt to describe the phenomena related to the advent of coherent and nonlinear lightwave communications. The fundamental aspects of these phenomena are emphasized, and possible new avenues for the practical achievement of theoretically predicted limits are examined.

The book is addressed to research telecommunications engineers dealing with optical communications systems, to engineers dealing with broadband telecommunications systems, to postgraduate electrical engineering students, and particularly to senior-level electrical engineering students. The background knowledge necessary to study this book is that of typical senior-level engineering students. It is assumed that the reader is familiar with electromagnetic theory, basic concepts of optics and electronics, elementary differential equations, and fundamental principles of communications theory. The book is organized to give a clear and logical sequence of topics, with necessary explanations in the Appendixes.

Acknowledgments

I owe gratitude to Prof. B. Stanic and Prof. G. Petrovic for critical proofreading of some chapters of the manuscript. I am also indebted to my friends D. Vucic, B. Radenovic, and S. Sretenov, whose suggestions enhanced the content and organization of the book.

I am very grateful to the scientists and engineers whose contributions have appeared in the open literature and which were the basic source material for this book.

My personal thanks to Dave McCarthy for his moral support and encouragement.

Finally, I am grateful to my wife, Rada, and my daughters, Neda and Marija, for their patience and encouragement during the time I devoted to writing this book.

Chapter 1

Introduction

1.1 THE MEANING AND IMPORTANCE OF COHERENT AND NONLINEAR LIGHTWAVE COMMUNICATIONS

It is well known that the discovery of laser radiation in 1960 [1] stimulated the big movement in the development of optical communications. Since the efficient source of light signals with relatively high space and time coherency was obtained, light radiation has possessed a small space divergency and narrow spectral linewidth. Thus, the necessary conditions for efficient transmission of great amounts of information over long distances were created. That is the reason why space and time coherency of light radiation were pointed out, and efforts were made for their maximal usage [2, 3].

At the very beginning, great care was devoted to the realization of laser transmission through the free space, or atmosphere. Laser atmosphere transmission should have been both a replacement for and an addition to radio-relay links, with improved transmission capacity. But the first theoretical works announced that the influence of atmosphere on the transmission quality was rather high, which practical experiments later confirmed. To achieve an as high as possible signal-to-noise (SNR) ratio, or rather the minimum bit-error rate (BER) a few lightwave modulation methods were proposed. *Intensity modulation* (IM) was the simplest method, but the feature of laser radiation coherency was not used entirely. This modulation method was based on linear variation of lightwave intensity with amplitude of modulating electrical signal. During that process care was not taken about the phase of the carrier wave, so the spectral linewidth of the carrier optical signal was much higher than spreading due to modulation. The detection of such a modulated signal was direct, so that detection was called *direct detection* (DD). In DD, the information is available in the baseband of frequencies instantaneously with optoelectronic signal conversion in a receiver photodetector. Hence, there is no conversion of optical frequencies by use of a local optical source. The noise influence on the quality of atmosphere transmission for IM signals was rather high, and it deteriorated the transmission capacity and decreased transmission line length.

When it became clear that, similar to transmission of radio signals, angle modulation methods offer significant advantages over IM, these methods became the subject of theoretical and practical investigations. An angle modulation is a variation of angular parameters (frequency and phase) of the carrier wave, assuming that the lightwave has the coherent nature. Such a modulation of lightwave increases the SNR at the receiver end, which has the maximum value when a local optical source is used in the detection scheme (heterodyne and homodyne detection) [4]. Compared to DD, detection with a local optical source is more complicated for practical purposes, because precise control of carrier wave parameters and of local source wave parameters is required.

Laser atmosphere transmission did not give expected advantages because of the very complex realization of the lightwave modulator and lightwave demodulator, on one side, and the great negative influence of atmosphere turbulence, on the other side [2, 5].

This fact directed research to the transmission of information by optical fibers (or rather lightguides). The production of efficient semiconductor light sources (lasers and light-emitting diodes) and low-loss broadband optical fibers meant final rejection of laser atmosphere transmission and the turning toward optical fiber transmission. Since the transmission of optical-fiber signals is accompanied by a relatively small influence of the optical source and the optical-fiber noise, the application of IM and DD schemes was rather efficient and relatively simple for a realization. All that instigated immense development in lightwave telecommunications. But the term *lightwave telecommunications* actually referred only to IM/DD systems [6, 7]. Such a transmission shows all its advantages through experimental verification and practical application, so we cannot say that the difficulties in the realizations of IM/DD lightwave systems led to an idea about coherent lightwave systems.

The idea of the application of coherent detection schemes in lightwave transmission systems came about, rather, by the sense that there is "a lot of space" for improvement of optical transmission quality [8–13]. That "empty space" means the entire use of coherent features of light, or the practical employment of coherent lightwave communications. The term *coherent lightwave communications* was introduced in 1981, after a few distinguished theoretical and practical results, and clearly signalled that the era of such communications had begun. The largest contributions to such a state were the production of single-mode semiconductor lasers with very narrow spectral linewidth and stabilized carrier frequency, realization of efficient electro-optical modulators and demodulators in an integrated optoelectronic version, and efficient methods for control of the polarization state of the incoming light signal.

Although there is some unclearness about the meaning of the term "coherent lightwave communications," it is assumed that coherent lightwave systems include all lightwave systems where a detection scheme with a local optical oscillator is applied. Thus, some realizations with amplitude modulation are included in coherent communications, as well. Hence, the subjects considered here are all systems with frequency and phase modulations of optical signal (the detection method is not important) and all systems that employ a local optical oscillator detection scheme, regardless of the kind of applied modulation. Such an approach is applied in this book, with selective analysis of relevant problems according to their practical importance.

As for the term *nonlinear lightwave communications,* the possible employment of nonlinear effects was announced before 1980 [14, 15], but this term dates to 1980–1981, when it was shown that the nonlinear effects, appearing in the optical fibers, can be used to improve characteristics of existing IM/DD lightwave systems [16, 17]. The influence of nonlinear effects can be observed when a relatively large level of optical power is being injected into an optical fiber. Some of these effects, such as self-phase modulation, have unfavorable influence on the entire characteristics of lightwave transmission system, while others (specifically, soliton generation regime and stimulated Raman scattering) can be efficiently used to increase the transmission capacity of lightwave IM/DD systems. The soliton propagation regime in optical fiber is induced by mutual compensation of the dispersion effect by the nonlinear self-phase modulation effect, leading to a considerable increase of system transmission capacity. The stimulated Raman scattering can be employed for a realization of efficient optical amplifiers for nonlinear systems.

Hence, the term "nonlinear lightwave communications" refers to IM/DD systems with predominant favorable influence of nonlinear effects on processes of light generation and light propagation through optical fibers. Such a system is one with achieved soliton regime of generation and propagation of the light signal. The subjects that are considered in nonlinear lightwave telecommunications are other nonlinear effects, both with favorable and unfavorable influence on the system characteristics, with the aim to suppress the unfavorable effects as much as possible and employ the favorable effects wherever possible [18].

The first impression may be that there is no close relation between coherent and nonlinear systems, but that is not quite true. The nonlinear effects can strongly influence the coherent system characteristics, and, vice versa, the coherence features of the light signal determine the character of nonlinear effects [19–20].

It is clear that coherent and nonlinear communications present two directions in the future development and applications of lightwave communications systems [21–32]. We can freely say that 1980–1981 began a new, second era in optical communications development, which will not be over in the near future. This second era should lead toward the realization of lightwave systems whose characteristics will approach the theoretically prescribed limits. But all of this does not mean that the attention to classical IM/DD lightwave systems will decrease, because realization of such systems is simpler and less expensive. On the other hand, IM/DD systems offer a lot of possibilities and a long period of wide applications.

1.2 THE STRUCTURE AND MAIN ADVANTAGES OF COHERENT LIGHTWAVE SYSTEMS

The meaning of the term coherent lightwave communications, which will be used in the next considerations, has been explained. Since the subject of our interest is, above all, digital transmission systems, we will consider the characteristics of digital systems for transmission of amplitude-modulated signals, frequency-modulated signals and phase-modulated signals. These signals are known as amplitude-shift keying (ASK), frequency-shift keying (FSK), and phase-shift keying (PSK) signals. (In this book, we will use the acronyms to denote not only the signals but the corresponding transmission systems, as well.)

The general scheme of a coherent-lightwave system we are considering is shown in Figure 1.1. For the optical transmitter light source, a semiconductor laser with

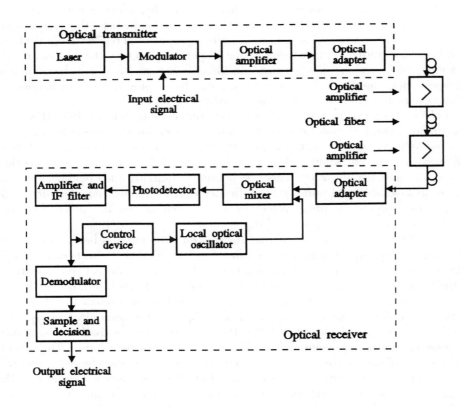

Figure 1.1 The basic scheme of a coherent lightwave system.

stabilized parameters of output optical signal is used. Modulation of the optical signal can be either internal or external. In the internal modulation scheme, the direct current signal injected into the laser structure modulates output optical radiation, while the external modulation scheme includes application of an additional modulator, to exert the influence on passing referent optical signals. The external modulation process is based on some physical effects, such as electro-optical effect, magneto-optical effect, and acousto-optical effect. The modulated signal can be amplified if necessary, to compensate both the optical power loss in the modulator and the coupling losses. A few types of optical amplifiers can be used for this purpose; their characteristics will be discussed in Chapter 8.

The optical adapter in the optical transmitter at the transmitting side has a double role: (1) to provide optimal coupling of the optical source signal and the fundamental HE_{11} mode in the single-mode optical fiber, and (2) to match the polarizations of the light source signal and HE_{11} mode (or, rather, to adapt the polarization state of the light source signal to one of two polarization eigenstates in signal-mode optical fibers).

Transmission of the angularly modulated optical signals between the optical transmitter and the optical receiver is performed by a single-mode polarization-maintaining optical fiber. As an alternative, the standard single-mode optical fiber can be used, but in combination with some polarization controllers employed before the photodetector or in combination with a phase/polarization-diversity detection scheme. Optical-fiber transmission lines can include a necessary number of optical amplifiers, which compensate for the optical power loss in the optical fiber.

The receiver optical adapter plays the inverse role of that in the optical transmitter. In the optical mixer, the information optical signal and the optical signal from a local optical oscillator are received, and their efficient mixing is made. The detection efficiency of mixed optical signal depends on the time and space coherency of both the incoming information optical signal and the local oscillator optical signal. The detection process can be either heterodyne or homodyne in nature. In the former case, the carrier frequency, ω_s, of incoming the optical signal and the frequency, ω_L, of the local optical oscillator signal differ for the value $\delta\omega = |\omega_s - \omega_L|$, while the homodyne detection is characterized by $\delta\omega = 0$.

The combined optical signal from the optical mixer falls to the photodetector with sufficiently large frequency bandwidth, where the optical signal is being converted into an electrical one, which carries the transmitted information. The electrical signal after filtration is led to the demodulator or to the decoder, where the sent information is extracted. In the homodyne detection process, the detected optical signal is transformed into an electrical signal from the frequency baseband region, and only a decoding operation can be made. The electrical signals from the photodetector in a heterodyne detection scheme are within the intermediate frequency (IF) range; hence, its demodulation is necessary before the decoding process. The demodulation is commonly imposed by a local electrical oscillator signal with the frequency equal to the central frequency of the IF-band filter (commonly called just an IF filter). Thus, there can be two local oscillators in a heterodyne detection scheme: the optical local oscillator and the electrical one. We use the full name for the corresponding oscillator to prevent possible confusion.

By the coherent detection scheme, according to Figure 1.1, with the proper choice of the local optical oscillator, the influence of a nonquantun nature noise can be suppressed. That possibility causes an increase in the sensitivity of a coherent optical receiver, in comparison with the sensitivity of a direct-detection optical receiver. The main difficulty in the coherent detection scheme realization is related to the necessity of adjustment between the optical information signal parameters and the local optical oscillator signal parameters. Random fluctuations of the information signal phase in the PSK modulation scheme can cause the "covering" of the information, while the random shifts of the carrier frequency in the FSK modulation scheme can lead to a strong decrease in signal-to-noise ratio. To achieve the stable relationships between the phases and the frequencies of the information optical signal and the local oscillator optical signal, efficient regulation schemes must be employed. These schemes are represented by the control device in the receiver feedback in Figure 1.1. The application of phase-locked loop (PLL) is the most efficient way for this regulation, so such a loop is the most commonly applied. The most complicated realization includes an optical PLL, which is the most efficient, especially if it is realized by integrated optoelectronic elements.

The most important advantage of a coherent lightwave system in comparison with an IM/DD lightwave system is the significant improvement of receiver sensitivity. That improvement is measured by the significant decrease of received optical power necessary to achieve the prescribed bit-error rate (usually less than 10^{-9}). The sensitivity of a coherent optical receiver is, in fact, determined by the level of quantum (shot) noise, while both the quantum noise and the thermal noise determine the sensitivity of a direct-detection optical receiver. Such a situation is caused by the influence of the local optical oscillator and suppression of the thermal noise in the coherent optical receiver.

The quality of the coherent lightwave transmission is measured by a signal-to-noise ratio increase (or receiver sensitivity improvement) in comparison with the signal-to-noise ratio of an ideal optical receiver. A signal-to-noise ratio only a few decibels below the value that corresponds to an ideal optical receiver can be reached in an IM/DD system, but only in the 0.8- to 0.9-μm wavelength region, with a silicium avalanche photodiode as a photodetector. Operation in the wavelength regions around 1.3 μm and 1.55 μm causes deterioration of the direct-detection receiver sensitivity due to the stronger influence of photodiode shot noise than in the 0.8- to 0.9-μm wavelength region [7].

The sensitivity improvement of the coherent optical receiver is determined by employment of the heterodyne (or homodyne) detection scheme and by the use of synchronous modulation methods. Thus, all types of coherent lightwave systems possess improved receiver sensitivity in comparison with direct detection optical receivers, but the extent of that improvement depends on both the applied detection scheme and the modulation type.

The heterodyne and the homodyne detections are the only realizable schemes that approach the receiver sensitivity of an ideal optical receiver in the most attactive wavelength regions (around 1.3 μm and 1.55 μm) [9, 10]. The improvement in coherent-receiver sensitivity can be more effectively illustrated by comparing different modulation/detection schemes with the IM/DD lightwave system, taking the direct-detection receiver sensitivity as a referent. Thus, the improvement of receiver sensitivity in an ASK/heterodyne detection scheme is 10–25 dB compared with the referent receiver sensitivity (IM/DD receiver sensitivity). Further improvement can be reached by application of frequency- or phase-modulation schemes instead of the amplitude one. The application of the FSK method brings the sensitivity benefit of 3 dB compared with the ASK scheme, while the PSK-modulation scheme provides further improvement of 3 dB compared with the FSK modulation scheme. Still further cumulative improvement can be obtained by PSK modulation and homodyne detection. It is possible to reach another cumulative improvement of the receiver sensitivity by employment of the multivalued modulation schemes, such as 4-valued PSK and 8-valued PSK, which bring an additional improvement of 3 dB and 5 dB, respectively, compared with the mentioned 2-valued PSK modulation scheme. All this will be considered in detail in Chapter 2.

It is interesting to express the optical receiver sensitivities as the necessary number of photons in the optical signal for a prescribed BER. Thus, the most sensitive direct-detection optical receiver needs about 700 photons per bit, a binary PSK/homodyne scheme needs about 18 photons per bit, while a 32-valued FSK/heterodyne scheme needs only 1.4 photons per bit. At the same time, the theoretical minimum for a coherent optical system is 0.02 photons per bit, while the minimum for an IM/DD system is 19 photons per bit [9].

The second large advantage of coherent lightwave systems, compared with the IM/DD systems, is related to the higher frequency selectivity, because it is made by an IF amplifier (an IF filter) instead of an optical filter. In such a way, very precise wavelength division multiplexing (WDM) can be performed with a small spectral distance between the individual carrier frequencies. Because of this fact, the term *frequency division multiplex* (FDM) is used rather than *wavelength division multiplex*. Thus, the very efficient filling of the spectral regions around wavelength 1.3 μm and 1.55 μm with the lot number of independent optical channels is provided. It will play a great role in different types of broadband communication networks.

The realization of FDM coherent optical systems is accompanied by additional losses in the system, appearing at the coupling points and in the multiplexing or demultiplexing elements. These additional losses could be, in general, compensated for by an optimal number of inserted optical amplifiers. In such a way, the efficient postmodulation amplification of the multiplexed optical signals can be performed by a laser amplifier, or the efficient line amplification of optical signals can be made by the erbium-doped optical fiber amplifiers, for example.

Another advantage of coherent lightwave systems is related to the further spreading of modulation-frequency bandwidth of the light source. That is, the modulation bandwidth of the semiconductor laser in a coherent optical transmitter is not limited by the laser resonant frequency, as was the case in an intensity-modulated transmitter. Further spreading of the modulation bandwidth in an angular modulation scheme can be performed by the employment of an external optical modulator. The advantages, in reference to the practical aspects of system design, will be discussed in Chapter 3 and Chapter 4.

1.3 THE STRUCTURE AND MAIN ADVANTAGES OF NONLINEAR LIGHTWAVE SYSTEMS

The general scheme of a nonlinear lightwave system is shown in Figure 1.2. It can be seen that this scheme is basically the same as the general scheme of an IM/DD system [6], because there are the same elements that appear in a linear optical transmission system. The main difference is related to the operating regime of the light source in the optical transmitter, which must generate the optical pulses with sufficient optical power and strongly defined shape. To generate and preserve the soliton regime in a nonlinear lightwave system, the optical pulses must have the shape defined by hyperbolic secant function, with the total energy per pulse above some critical value [14, 17].

As for the term *nonlinear lightwave communications*, some explanations are necessary. It has been mentioned that nonlinear effects can impose either favorable

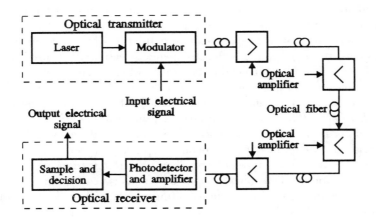

Figure 1.2 The general scheme of a nonlinear lightwave system.

or unfavorable influences on the entire transmission characteristics of the lightwave system. Of course, the favorable effects should be used to the largest possible extent, while the unfavorable effects are to be suppressed as much as possible. The term *nonlinear lightwave system* is related to the system shown in Figure 1.2, where the regimes of soliton generation and soliton propagation through an optical fiber have been achieved. The amplification of solitons in such a case actually means their reshaping by erbium-doped fiber amplifiers, above all. The accent on the application of erbium-doped amplifiers is based on the fact that, at this time, they are the most efficient optical amplifiers for such a purpose.

On the other hand, this book will consider other nonlinear effects, as well, in order to obtain the complete representation about nonlinear lightwave systems. These effects will be considered in detail in Chapter 6; now we are giving only a brief review of the nonlinear effects and the conditions for their appearance. The main nonlinear effects that can appear in optical fibers are: stimulated Raman scattering, stimulated Brillouin scattering, self-phase modulation, and four-photon mixing process (or three wave mixing) [16, 19, 20].

Raman-scattering processes in an optical fiber can have both a spontaneous and a stimulated nature. Spontaneous Raman scattering appears as a consequence of the interaction of an incident optical wave and the molecular vibrations in the material (glass), where the incident photons cause the stimulated transitions of electrons and the generation of new photons with higher or lower wavelengths. In such a process, a highly dumped vibrational wave is generated as well, as a consequence of the energy adjustment between the incident and the secondary generated photons. If the secondary generated photons possess a higher wavelength than the incident ones, they form so-called *Stokes radiation* or *Stokes line*; if they have a lower wavelength, an anti-Stokes line is generated. For stimulated Raman scattering to occur, the power of the incident optical signal should be above some critical value. In such a case, the scattered optical signal takes the form of laser radiation with corresponding coherency characteristics. The critical value of incident optical power was evaluated in a few papers and depends on the optical fiber characteristics and the incident optical signal wavelength. But, as an illustration, we can say that the critical power for the stimulated Raman scattering in a standard single-mode optical fiber at a wavelength of 1.5 μm is about 1.7W for single-channel transmission [16].

Hence, the stimulated Raman scattering is the frequency conversion process, where the stimulated generation of the light can be achieved. This effect can be efficiently used in the design of Raman lasers and Raman optical amplifiers. Raman amplifiers can be efficiently used in both coherent and nonlinear lightwave communication systems, especially for amplification of FDM optical signals, due to their broadband characteristics. But the stimulated Raman scattering process can lead to unwanted crosstalking between the optical channels in WDM and FDM lightwave systems, and it must be taken into account in the process of system design [20].

Stimulated Brillouin scattering occurs for considerably smaller incident optical peak powers (about 5–35 mW for single-channel transmission). The stimulated Brillouin scattering arises as a consequence of interaction between the incident optical wave and the acoustic waves, when the corresponding wave vectors are matched. Because of this matching, the stimulated Brillouin scattering may have only the backward character (while the stimulated Raman scattering may have both the forward and the backward characters). The frequency shift of the scattered signal depends on the wavelength of the incident optical signal, the refractive index of the optical fiber core, and the velocity of the acoustic waves. At the same time, the frequency bandwidth of the scattered signal is considerably narrower than the bandwidth of the Raman scattered signal, and that limits the eventual application of Brillouin amplifiers to low bit-rate systems. It should be mentioned that the stimulated Brillouin scattering, due to its own character, can result in an excess attenuation of short-wavelength channels in a WDM system.

Four-photon mixing, or three-wave mixing, is the parametric nonlinear process that appears as a consequence of the mixing of an incident optical wave with signals that have arisen due to either stimulated Raman scattering or stimulated Brillouin scattering. The newly generated optical signals have both the Stokes nature and the anti-Stokes nature. The four-photon process is strongly determined by the conditions of phase matching between the incident and the scattered signals. Since the phase-matching conditions are rather strong, the four-photon process does not, in general, occur as an independent process, but rather it accompanies the Raman or Brillouin stimulated scattering processes. Accordingly, the unfavorable influence of four-photon mixing on the lightwave system characteristics can be expected, above all, in multichannel systems.

Self-phase modulation is the nonlinear process that causes the spreading of the spectrum of input optical signals during the signal propagation through an optical fiber. The spectrum spreading depends on the optical-signal intensity and can be understood as a differential phase shift of carrier optical frequency between the center and the edges of an optical pulse. The frequency shift is caused by a nonlinear dependence of refractive index in the optical signal level. Hence, the difference between the levels, corresponding to the center and the edges of an optical pulse, induces the corresponding phase shift. The induced phase shift is rather small for a short distance, but it can be relatively large for large optical-fiber lengths. The self-phase modulation effect can occur at the optical power levels below 50 mW, so its influence should be taken into account often. The self-phase modulation effect, in general, leads to deterioration of transmission quality in a long-length lightwave system, but the self-phase modulation can play a positive role when the special operating regime is reached. Such a regime is, in fact, the solitons regime of the generation and propagation of optical pulses.

The soliton regime in an optical fiber can occur if the self-phase effect is properly compensated for by a dispersion one. Such a case does happen in the wavelength region corresponding to the so-called *anomalous dispersion* in an optical fiber. The optical solitons, or the optical pulses with characteristic shape, can propagate through an optical fiber with preserving shape. To achieve the solitons regime, the light source should be modulated by the pulses of strongly defined shape (usually the hyperbolic secant type shape). The peak power necessary for solitons generation depends on the geometric characteristics of the optical fiber cross-section, the carrier wavelength, the pulse width, and the optical-fiber dispersion characteristics at the carrier wavelength. The proper time distance between the neighboring soliton pulses must be chosen during their generation process in an optical source, to prevent the mutual interaction between the neighboring pulses in the optical fiber [21].

It was shown that the hyperbolic secant is not the only temporal shape of the pulses that provides the stable solitons transmission [22]. Namely, it was shown that the input pulses with the Gaussian shape can lead to stable soliton transmission, with some improvement, which refers to the prevention of pulses mutual interacting. This is very important for practical realization of the optical transmitter from Figure 1.2, because the generation of a Gaussian-shape pulse does not present a problem.

The behavior of soliton pulses over the optical-fiber line does not depend on the dispersion characteristics, because the dispersion effect is compensated for by the nonlinear one, but it does depend on the optical-fiber loss at the carrier optical wavelength. The optical-fiber loss is the cause of soliton pulses spreading during their propagation through the optical fiber. The spreading has the adiabatic nature, because the area under the soliton envelope stays unchanged. The soliton-pulse spreading causes the approaching of neighboring pulses, which can enter into the area of their mutual interaction. The mutual interaction of the neighboring soliton pulses must be prevented, by the periodical insertion of optical amplifiers into optical fiber line, as shown in Figure 1.2. The main role of the optical amplifiers is reshaping of the soliton pulses or, in fact, the narrowing of the pulse width and the increase of pulse amplitude. The amplification process should not destroy the soliton regime, so the amplification coefficient is chosen to be slightly above the unity.

Several methods of soliton reshaping have been proposed so far. Employment of erbium-doped optical-fiber amplifiers is the most efficient method that has been proposed and verified [21], but the soliton reshaping by periodically inserted Raman amplifiers can be also done [22].

Other elements from Figure 1.2, belonging to a nonlinear lightwave system, are practically the same as those used in high-speed IM/DD lightwave systems. Thus, very fast photodetectors and electronic circuits should be used in such a system. The optical coupling elements in a nonlinear lightwave system must have losses as low as possible, because high coupling losses can destroy the soliton regime.

The main advantages of nonlinear lightwave systems over classical IM/DD systems can be easily understood by comparing their transmission capacities. The transmission rate and the transmission capacity of a linear IM/DD lightwave system are determined by the dispersion characteristics of the optical fibers. The extent of the dispersion influence can be measured by the extent of pulse spreading in the optical fiber. Thus, if the initial pulse width in the single-mode optical fiber is equal to $2t_0$, which corresponds to the frequency bandwidth $(t_0)^{-1}$, it will be doubled after a distance proportional to $(t_0)^2$. Practically speaking, that means that the 10-ps-wide optical pulse doubles its width after about 650m, propagating through standard single-mode optical fiber at 1.5-μm wavelength. The propagation in the "zero dispersion region" around 1.3 μm, where the total dispersion in a silica-based optical fiber has the absolute dispersion minimum, leads to the pulse width doubling after a distance proportional to $(t_0)^3$, which is the minimal pulse spreading that can be achieved in a linear lightwave transmission system. The more exact calculation [24] shows that the minimal pulse width at the input end of a single-mode optical fiber and maximal transmission distance, L, in the zero dispersion wavelength region are related with the expression $2t_0L^{-1/3} = 1.4$ ps(km)$^{-1/3}$. Hence, for example, the transmission with a signal bit-rate of 130 Gb/s over the length of about 20 km can be performed. Such a transmission can be realized *only* if the operating wavelength is equal to the zero dispersion wavelength. However, any small operating wavelength shift from the zero dispersion value causes drastic limitation of the transmitting bit rate. Thus, a wavelength shift of only 1% decreases the mentioned bit rate to a value of 10 Gb/s.

Nonlinear transmission under the same conditions provides a doubling of the pulse width after 15 km (instead of 650m). Hence, the transmission capacity of a nonlinear transmission system is about 100 times larger than the transmission capacity of the optimized linear transmission. The operational frequency shift from the prescribed value is not so critical any more, which is an additional advantage of nonlinear transmission. Finally, we can add that nonlinear optical pulses can be more efficiently regenerated by an optical amplifier than can linear ones. All these advantages are related to the high-bit-rate, long-length lightwave transmission systems, which are going to be widely used in the near future.

1.4 PROBLEMS IN COHERENT AND NONLINEAR LIGHTWAVE SYSTEMS

There are a few serious practical problems that should be solved before the wide practical application of coherent and nonlinear lightwave systems. Resolution of these problems will determine the direction of the development and optimization processes in the areas of coherent and nonlinear lightwave communications.

The most important technical problems that should be adequately solved in a coherent lightwave system are related mainly to further improvement in the stabilization of the laser operating regime (frequency stabilization, noise suppression, wider wavelength tuning range). Also, suppression of the influence of nonlinear effects and realization of efficient integrated optoelectronic devices for modulation/demodulation functions play an essential role. At the same time, the soliton lasers and optical amplifiers for soliton reshaping need further improvement before nonlinear lightwave systems leave the experimental stage.

1.5 ORGANIZATION OF THIS BOOK

This book consists of 10 chapters, which cover particularly closed themes. The first part of the book is related to coherent lightwave telecommunications, while the second part considers nonlinear lightwave systems. Some parts, such as Chapter 8, are relevant for both coherent and nonlinear lightwave systems.

Chapter 2 deals with coherent modulation/demodulation schemes and evaluates coherent receiver sensitivities for ASK, FSK, PSK, and differential phase-shift keying (DPSK) detection schemes.

Chapter 3 describes the methods for internal and external modulations of the lasers, with emphasis on semiconductor laser characteristics. Chapter 4 is devoted to the relevant types of coherent optical receivers, with special care paid to PLL application. Chapter 5 deals with coherent optical communication systems, with special attention to FDM systems.

Dominant nonlinear effects in optical fibers and the generation and transmission of optical solitons are described in Chapter 6 and Chapter 7. Chapter 8 deals with the characteristics of optical amplifiers used in both coherent and nonlinear systems. Chapter 9 discusses realization of nonlinear communication systems and evaluates bit-error rates in nonlinear systems. Finally, the Appendixes review some relevant explanations and mathematical relations.

REFERENCES

[1] Maiman, T. H., "Stimulated Optical Radiation in Ruby," *Nature* (London), 6(1960), pp. 106–118.
[2] Creamer, A. R., "Free-Space Optical Communications," *Signal*, 32(1977), pp. 26–32.
[3] Miller, S. E., et al., "Research Toward Optical Fiber Transmission Systems," *IEEE Proc.*, 61(1973), pp. 1703–1751.
[4] Oliver, B. M., "Signal to Noise Ratios in Photoelectric Mixing," *IRE Proc.*, 49(1961), pp. 1960–1961.
[5] Gordon, J. P., "Quantum Effects in Communication Systems," *IRE Proc.*, 50(1962), pp. 1898–1908.
[6] Howes, M. J., and D. V. Morgan, *Optical Fibre Communications*, New York: Willey, 1980.
[7] Keisher, G., *Optical Fiber Communications*, Tokyo: McGraw Hill, 1984.
[8] Yamamoto, Y., "Receiver Performance Evaluation of Various Digital Optical Modulation-Demodulation Systems in the 0.5-10 μm Wavelength Region," *IEEE J. Quantum Electronics*, QE-16(1980), pp. 1251–1259.
[9] Yamamoto, Y., and T. Kirmura, "Coherent Optical Fiber Transmission Systems," *IEEE J. Quantum Electronics*, QE-17(1981), pp. 919–935.
[10] Favre, F., et al., "Progress Towards Heterodyne-Type Single Mode Fiber Communication Systems," *IEEE J. Quantum Electron.*, QE-17(1981), pp. 897–906.
[11] Okoshi, T., and K. Kikuchi, "Heterodyne-Type Optical Fiber Communications," *J. Opt. Commun.*, 2(1981), pp. 82–88.

[12] Okoshi, T., "Heterodyne and Coherent Optical Fiber Communications: Recent Progress," *IEEE Trans. Microwave Theory Techn.*, MTT-30(1982), pp. 1138–1149.

[13] Okoshi, T., and K. Kikuchi, *Coherent Optical Fiber Communications*, Tokyo: Kluwer Academic Publishers, 1988.

[14] Hasegawa, A., and F. Tapert, "Transmission of Stationary Nonlinear Optical Pulses in Dispersive Dielectric Fibers," *Appl. Phys. Lett.*, 23(1973), pp. 142–144.

[15] Bloom, D. M., et al., "Direct Demonstration of Distortionless Picosecond Pulse Propagation in Kilometer Length Optical Fibers," *Opt. Lett.*, 4(1979), pp. 297–299.

[16] Stolen, R. H., "Nonlinearity in Fiber Transmission," *IEEE Proc.*, 68(1980), pp. 1232–1235.

[17] Hasegawa, A., and Y. Kodama, "Signal Transmission by Optical Solitons in Monomode Fiber," *IEEE Proc.*, 69(1981), pp. 1145–1150.

[18] Kodama, Y., and A. Hasegawa, "Generation of Asymptotically Stable Optical Solitons and Suppression of the Gordon Hans Effect," *Optics Letters*, 17(1992), pp. 31–33.

[19] Stolen, R., and J. E. Bjorkholm, "Parametric Amplification and Frequency Conversion in Optical Fibers," *IEEE J. Quant. Electron.*, QE-18(1982), pp. 1062–1071.

[20] Chraplyvy, A. R., "Limitations of Lightwave Communications Imposed by Optical-Fiber Nonlinearities," *IEEE/OSA J. Lightwave Techn.*, LT-8(1990), pp. 1548–1557.

[21] Taga, H., et al., "Multi-Thousand Kilometer Optical Soliton Data Transmission Experiments at 5 Gb/s Using an Electroabsorption Modulator Pulse Generator," *IEEE/OSA J. Lightwave Techn.*, LT-12(1994), pp. 231–235.

[22] Hasegawa, A., "Amplification and Reshaping of Optical Soliton in a Glass Fiber-IV: Use of the Stimulated Raman Process," *Opt. Lett.*, 8(1983), pp. 650–652.

[23] Mollenauer, L. F., and K. Smith, "Demonstration of Soliton Transmission Over More Than 4000 km in Fiber With Loss Periodically Compensated by Raman Gain," *Optics Lett.*, 13(1988), pp. 675–677.

[24] Marcuse, D., and C. Lin, "Low Dispersion Single-Mode Fiber Transmission—The Question of Practical Versus Theoretical Maximum Transmission Bandwidth," *IEEE J. Quantum Electron.*, QE-17(1981), pp. 869–874.

[25] Iannone, E., et al., "High Speed DPSK Coherent Systems in the Presence of Chromatic Dispersion and Kerr Effect," *IEEE/OSA J. Lightwave Techn.*, LT-11(1993), pp. 1478–1485.

[26] Naito, T., et al., "Optimum System Parameters for Multigigabit CPFSK Optical Heterodyne Detection System," *IEEE/OSA J. Lightwave Techn.*, LT-12(1994), pp. 1835–1841.

[27] Chandrakumar et al., "Combination of In-Line Filtering and Receiver Dispersion Compensation for Optimized Soliton Transmission," *IEEE/OSA J. Lightwave Techn.*, LT-12(1994), pp. 1047–1051.

[28] Yamazaki, S., et al., "Compensation of Chromatic Dispersion and Nonlinear Effects in High Dispersive Coherent Optical Repeater Transmission System," *IEEE/OSA J. Lightwave Techn.*, LT-11(1993), pp. 603–611.

[29] Suzuki, N., et al., "Simultaneous Compensation of Laser Chirp, Kerr Effect and Dispersion in 10 Gb/s Long Haul Transmission Systems," *IEEE/OSA J. Lightwave Techn.*, LT-11(1993), pp. 1486–1494.

[30] Taga, H., et al., "Long Distance Multichannel WDM Transmission Experiments Using Er-Doped Fiber Amplifiers," *IEEE/OSA J. Lightwave Techn.*, LT-12(1994), pp. 1448–1453.

[31] Nakazava, M., et al., "10 Gbit/s 1200 km Error Free Soliton Data Transmission Using Erbium Doped Fibre Amplifiers," *Electron. Letters*, 28(1992), pp. 817–818.

[32] Mollenauer, L. F., et al., "Demonstration of Error Free Soliton Transmission Over More Than 15000 km at 5 Gbit/s, Single Channel, and Over 11000 km at 10 Gbit/s in Two Channel WDM," *Electron. Letters*, 28(1992), pp. 792–794.

Chapter 2

Coherent Optical Receiver Sensitivity

2.1 INTRODUCTION

It was pointed out in Chapter 1 that the principal advantage of coherent lightwave systems over IM/DD systems is the increase in optical receiver sensitivity. This increase is caused mainly by the suppression of the thermal noise influence in the optical receiver; thus, the quantum noise becomes dominant. Further improvement, obtained by application of angular modulation of the digital signals (which is more beneficial than the amplitude-modulation method), is well known from the general theory of telecommunications [1].

Optical receiver sensitivity is defined as the minimal optical power necessary at the receiving end to reach the prescribed error probability, or bit-error rate (BER). The prescribed value of BER for digital lightwave systems is usually 10^{-9}. In this chapter, the error probabilities will be evaluated for different modulation/detection schemes, and the corresponding quantitative relations dealing with receiver sensitivities will be established. In the evaluation of coherent receiver sensitivity, we will use the corresponding relations, which refer to the IM/DD lightwave systems, given in Appendixes E, F, and G.

This chapter considers the basic relations of detection with a local optical oscillator (heterodyne and homodyne detection) and analyzes the most important modulation/detection schemes.

2.2 DETECTION WITH A LOCAL OPTICAL OSCILLATOR

At the very beginning of the analysis of detection with use of local optical oscillator, we can assume that the electrical fields of the signal source and the local optical oscillator are ideally adjusted in the optical mixer. Hence, they can be represented in the form of normal plane waves propagating in the same direction and in the same space mode, as well. Because of that, the detection procedure can be considered only in time domain. In the general case of heterodyne detection, the optical field incoming from the optical fiber line can be expressed as

$$f_s(t) = \text{Re}[a_s(t)\exp(j\omega_s t)] \qquad (2.1)$$

where $a_s(t)$ is the complex envelope of the input optical field (either electrical or magnetic) belonging to the signal source, and ω_s is the optical frequency of the field.

The optical radiation of the local optical oscillator can be represented in the form of a plane wave with the frequency ω_L, or

$$f_L(t) = \text{Re}[a_L \exp(j\omega_L t)] \qquad (2.2)$$

The combined field at the output of optical mixer has the form

$$f(t) = \text{Re}[a_s(t)\exp(j\omega_s t) + a_L \exp(j\omega_L t)] \qquad (2.3)$$

The complex envelopes from (2.3) can be represented by the amplitude-phase functions, so that we have

$$f(t) = \text{Re}\{|a_s| \exp[j\omega_s t + \varphi_s(t)] + |a_L| \exp(j\omega_L t + \varphi_L)\} \qquad (2.4)$$

The intensity of the combined field at the area of photodetector is

$$I(t) = |f^2(t)| = |a_s(t)|^2 + |a_L|^2 + 2|a_s(t)||a_L(t)| \\
\cdot \cos[(\omega_s - \omega_L)t + \varphi_s(t) - \omega_L] \qquad (2.5)$$

The first term on the right side of (2.5) is the intensity of the signal field and presents the power of the incoming optical signal; the second and the third terms refer to the existence of the local oscillator radiation. The third term is the most interesting for further processing, because its amplitude is directly related to the amplitude of the signal field, while its phase is linearly changed with the variation of the signal field phase. Accordingly, the information contained in the amplitude or in the phase of the signal field is entirely preserved in the intensity of the combined field at the photodetector.

If it is assumed that the field intensity of the local optical oscillator is much higher than the intensity of the signal field, or when it is $|a_L|^2 \gg |a_s(t)|^2$, which is the most common case, (2.5) takes the form

$$I(t) \simeq |a_L|^2 + 2|a_s(t)||a_L|\cos[(\omega_s - \omega_L)t + \varphi_s(t) - \varphi_L] \qquad (2.6)$$

Hence, the useful information is only in the functions $|a_s(t)|$ and $|\varphi_s(t)|$ of the term, which refers to the product between the signal field and the local oscillator field. All intermodulation terms, appearing in other situations, are vanishing now. It is more suitable to express (2.6) by the corresponding optical powers. Since the power corresponds to the square of the amplitude, (2.6) can be rewritten in the form

$$P(t) \simeq P_L + 2\sqrt{P_L P_s(t)}\cos[\Delta\omega t + \Delta\varphi(t)] \tag{2.7}$$

where

$$\Delta\omega = \omega_s - \omega_L \text{ and } \Delta\varphi(t) = \varphi_s(t) - \varphi_L \tag{2.8}$$

It is obvious that the analytical expression for the optical power of the combined optical fields differs from the corresponding expression for direct detection. To perform reliable detection by the local optical oscillator, the stable relationships between the frequencies ω_s and ω_L and the phases φ_s and φ_L should be imposed. Since the frequency and the phase of the local oscillator field are directly summed (or subtracted) with the frequency and the phase of the signal field, any random variation of these parameters has influence on the level of the received optical field and cannot be separated from the "useful," or induced, variations due to modulation. Hence, the random variation must be eliminated or suppressed as much as possible, which is the most difficult task in the realization of coherent lightwave systems [2].

2.3 TOTAL NOISE IN DETECTION WITH A LOCAL OSCILLATOR

The noise characteristics in a coherent optical receiver will be considered in detail in Chapter 4. In this chapter we will use only the final expressions to make the proper evaluation of the error probability. The expressions that follow are the continuation of the relations that describe IM/DD systems, given in Appendix F.

The heterodyne detection scheme of the incoming optical signal is characterized by an additional term of the noise current variance, due to influence of the local oscillator optical power. The total variance of the noise current will depend on the sent binary state, so it will be higher for "1" state sent than for the "0" state sent. Besides, the equivalent bandwidth, B, of the noise is now equivalent to the doubled value of the noise bandwidth in the direct detection scheme. That is because heterodyne detection causes the frequency shift in the intermediate frequency region, and both the upper and the lower frequency sidebands should be taken into account. In a homodyne detection scheme, the equivalent noise bandwidth is equal to the equivalent noise bandwidth for a direct detection scheme, because homodyne detection provides the frequency shift to the baseband frequency region, where only the upper sideband should be taken into account. The noise in a general heterodyne detection scheme will be considered first.

The total variance of the noise current at the output of the photodetector when the "1" state is sent is equal to

$$\sigma_{h1}^2 = \overline{i_s^2} + \overline{i_d^2} + \overline{i_t^2} + \overline{i_L^2} \tag{2.9}$$

while the same variance for the transmitted the "0" state is

$$\sigma_{h0}^2 = i_d^2 + i_t^2 + i_L^2 \tag{2.10}$$

The terms on the right side of (2.9) and (2.10) have the following meanings:

- The variance of the quantum (or shot) noise due to the existence of an optical signal with power P_s is

$$\overline{i_s^2} = 2q\mathcal{R}P_s M^{x+2}(2B) \tag{2.11}$$

where q is the electron charge, $\mathcal{R}$ is the photodiode responsivity, M is the coefficient of avalanche amplification, and x is the coefficient of excess noise in the avalanche photodiode (APD). Coefficient M is equal to unity for PIN photodiodes, while its value is commonly in the region from 10 to 100 for APDs. The coefficient of excess noise, x, depends on the kind of photodiode (Appendix F). The photodiode responsivity is

$$\mathcal{R} = \frac{\eta q}{hf} \tag{2.12}$$

where η is the quantum efficiency of the photodiode, f is the carrier optical frequency, and h is the Planck's constant. The quantum efficiency of the photodiodes that are employed in optical receivers lies in the range from 0.5 to 0.85 [3]. Finally, B is the equivalent bandwidth of the noise (see Appendix B).
- The variance of the dark current noise is

$$\overline{i_d^2} = 2q\,(i_{ds} + i_{db}M^{x+2})(2B) \tag{2.13}$$

where i_{ds} denotes the surface dark current, while i_{db} denotes the bulk dark current, according to Appendix F.
- The variance of the thermal noise generated in the load resistance and the receiver amplifier, which depends on the amplifier design (see Chapter 4 and Appendix B), has the form

$$\overline{i_t^2} = \frac{4k\Theta}{R_L}(2B) \tag{2.14}$$

where k is Boltzmann's constant, Θ is the absolute temperature, and R_L is the load resistance.

• The variance of the quantum noise caused by the local oscillator optical power is

$$\overline{i_L^2} = 2q\Re P_L M^{x+2}(2B) \tag{2.15}$$

The power of the useful signal at the output of the (IF) filter can be found by knowing the corresponding signal current. The signal current at the output of the photodiode is found from (2.7) and has the form

$$I_{sL}(t) = 2\Re M\sqrt{P_s(t)P_L} \tag{2.16}$$

The SNR can be obtained from (2.9), (2.10), and (2.16), assuming that the powers are evaluated on the load resistance, R_L. Thus, we have

$$SNR_1 = \frac{R_L \, I_{sL}^2}{2 \, R_L \, \sigma_{h1}^2} = \frac{I_{sL}^2}{2 \, \sigma_{h1}^2} \quad \text{for "1" state} \tag{2.17}$$

$$SNR_0 = \frac{R_L \, I_{sL}^2}{2 \, R_L \, \sigma_{h0}^2} = \frac{I_{sL}^2}{2 \, \sigma_{h0}^2} \quad \text{for "0" state} \tag{2.18}$$

Since the variance of the quantum noise due to local oscillator influence is the dominant term in the expression for the total noise, the values SNR_1 and SNR_0 are, in fact, very close, so only one expression (for example, SNR_1) can be used in the next considerations. [The exact value of the SNR is given by (4.21)].

2.4 DIGITAL AMPLITUDE MODULATION OF OPTICAL SIGNALS

2.4.1 Modulation of Optical Signals

The binary amplitude-modulated optical signal (the ASK signal) in an optical transmitter is obtained by amplitude modulation of the carrier optical wave, $s_o(t)$, with the binary digital signal, $s_m(t)$, with the maximum value of the modulation depth. It is convenient to neglect the optical carrier linewidth at this moment and take it into account later on (in Subsection 2.4.5). The carrier optical wave with unity amplitude has the form

$$s_0(t) = \cos(\omega_s t + \varphi_s) \tag{2.19}$$

where $\omega_s = 2\pi f_s$ is the carrier frequency, and φ_s is the initial phase of the carrier optical wave. The carrier frequency is much larger than the modulation frequency, f_d (which is equal to $1/T_d$, where T_d is the digit period, or digit interval). The modulating digital electrical signal has the form

$$s_m(t) = \sum_{n=-\infty}^{\infty} a_n h_m(t - nT_d), \ a_n = 0 \text{ or } 1 \tag{2.20}$$

where h_m is the nonreturn to zero (NRZ) rectangular pulse with the amplitude A_m and the width T_d. The ASK signal at the output of optical transmitter has the form

$$s_A(t) = \begin{cases} h_m(t) \cos(\omega_s t + \varphi_s), & 0 \le t \le T_d \\ 0 & \text{for another } t \end{cases} \tag{2.21}$$

and is illustrated in Figure 2.1.

The spectrum of the signal in Figure 2.1 can be additionally shaped by the transmitter filter, in order to eliminate the high-order frequencies. The optical signal given by (2.21) propagates through the optical fiber line and reaches to photodetector.

2.4.2 Noncoherent Detection of Optical ASK Signals

Noncoherent detection in the optical receiver is the simplest detection method of ASK signals. In this case the detection is made by a local optical oscillator, but the receiver demodulator does not contain an electrical local oscillator. The current signal from the photodetector, with the amplitude given by (2.16), is amplified and passed

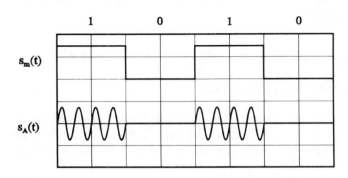

Figure 2.1 The time diagram of ASK signal.

through an IF filter with central frequency ω_c. Since the ASK spectrum consists of two sidebands, the IF filter must have a doubled bandwidth compared with the corresponding baseband filter in an IM/DD system. The detected envelope of the random electrical process, consisting of detected signals and narrowband noise, is sampled at the moments that correspond to the middle of digit intervals. After that the comparison with the chosen value of threshold b is performed, and the decision of what was sent ("1" or "0") is made.

The detected signal after the receiver filter can be presented by

$$s(t) = \begin{cases} Zh(t) \cos \omega_s t & \text{for state "1"} \\ 0 & \text{for state "0"} \end{cases} \tag{2.22}$$

where it is assumed that the initial phase is equal to zero. The parameter Z presents the transimpedance from the input current signal to demodulator output voltage, while $h(t)$ describes the wave shape obtained after the actions of the transfer functions of the optical fiber line, photodetector, and receiver amplifier on the wave form $h_m(t)$. If the central frequency of the IF filter is equal to the frequency shift between the incoming optical signal and the local oscillator optical signal, or if $\omega_c = \Delta\omega$, the further analysis corresponds to the case of the detection of narrowband signal and narrowband noise considered in Appendix D. This process is described by the expression

$$f_s(t) = s(t) + n(t) = Z\,[h(t) + x(t)]\cos \omega_c t - Z \sin \omega_c t \tag{2.23}$$

This expression can be normalized by division by Z, so the function $f(t) = f_s(t)/Z$ can be considered next. The envelope of the process $f(t)$ is

$$r(t) = \sqrt{[h(t) + x(t)]^2 + y(t)^2} \tag{2.24}$$

The instantaneous values of the envelope of function $f(t)$ are submitted to the Rice distribution, so we have

$$w_1(r) = \frac{r}{\sigma_1^2} \exp\left(-\frac{r^2 + S^2}{2\sigma_1^2}\right) I_0\left(\frac{rS}{\sigma_1^2}\right) \tag{2.25}$$

In (2.25) S denotes the amplitude of the current signal, while σ_1 is the standard deviation of the current noise for the case when the "1" state was sent. When the "0" state was transmitted, only the noise with average power σ_0^2 is present at the output of the IF filter. The instantaneous values of the envelopes of filtrated noise submit to the Rayleigh distribution (Appendix C); their probability density function is given as

$$w_0(r) = \frac{r}{\sigma_0^2} \exp\left(-\frac{r^2}{2\sigma_0^2}\right) \tag{2.26}$$

The sampled value of the envelope is compared with the threshold b, and the decision about the transmitted level is made. The total error probability can be found by the probability density functions $w_1(r)$ and $w_0(r)$, in the same way as has been done for IM/DD systems (Appendix E). The conditional probability that the "1" state will be chosen when the "0" state is sent is

$$P(1/0) = \int_b^\infty w_0(r)dr = \exp\left(-\frac{b}{2\sigma_0^2}\right) \tag{2.27}$$

The probability that the output from the decision circuit (or comparator) is "0" when the "1" state was sent is equal to

$$P(0/1) = \int_0^b w_1(r)dr \tag{2.28}$$

Equation (2.28) can be written in more suitable form as

$$P(0/1) = 1 - P(1/1) = 1 - Q\left(\frac{S}{\sigma_1}, \frac{b}{\sigma_1}\right) \tag{2.29}$$

where $Q(\alpha, \beta)$ is the so-called Marcum's Q-function [1], defined as

$$Q(\alpha, \beta) = \int_\beta^\infty tI_0(\alpha t) \exp\left(-\frac{t^2 + \alpha^2}{2}\right)dt \tag{2.30}$$

where I_0 is the Bessel function [4]. The Q-function is, in fact, the cumulative probability of the exact detection of the "1" state in a system with binary amplitude modulation. So, assuming that the probabilities of the appearance of the "0" and "1" states are equal in the pulse train, the final expression for error probability in an ASK system can be written in the form

$$P_{eASK,2} = 0.5\left\{\left[1 - Q\left[\frac{S}{\sigma_1}, \frac{b}{\sigma_1}\right]\right] + \exp\left(-\frac{b}{2\sigma_0}\right)\right\} \tag{2.31}$$

Equation (2.31) can be considered functionally dependent on the threshold value, because it is clear that for every S/σ_1 ratio there is some threshold value that minimizes the error probability. The value of the threshold increases with the increase of the S/σ_1 ratio.

The optimum threshold value is illustrated in Figure 2.2, which shows the Rice and Rayleigh distributions. The Rayleigh distribution has a maximum for $r = \sigma_0$, while the Rice distribution has the maximum at the point $r \simeq S$. If the threshold level is too small, then a level 1 will be registered in almost all cases; vice versa, too high a threshold means that a 0 level will be predominantly registered. Too small a threshold is denoted by b_1 in Figure 2.2, while b_2 denotes too high a threshold level. The optimum value of the threshold can be chosen in such a way that the conditional probabilities $P(0/1)$ and $P(1/0)$ are equal (the shadowed areas on the left and on the right from value b_{opt} are equal). To find b_{opt}, it is necessary to satisfy the conditions

$$\sigma_0^2 \simeq \sigma_1^2 = \sigma^2 \qquad (2.32)$$

and

$$I_0\left(\frac{Sb_{opt}}{\sigma^2}\right) \exp\left(-\frac{S^2}{2\sigma^2}\right) = 1 \qquad (2.33)$$

The condition in (2.32) is satisfied when the local optical oscillator has power considerably higher than the power of the incoming optical signal. It was shown that the approximation (2.32) causes the minimal error and that the obtained optimal threshold value is nearly equal to the exact optimal value. Condition (2.33) is satisfied if $b_{opt} = S/2$ [1, 2]. In that case, the Marcum's function becomes

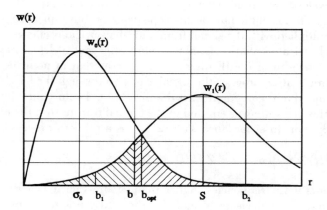

Figure 2.2 Determination of optimum threshold value in detection of ASK signals.

$$Q(\alpha, \beta) = 1 - 0.5 \; \text{erfc}\left(\frac{\alpha - \beta}{\sqrt{2}}\right) \tag{2.34}$$

By applying the approximation [4]

$$\text{erfc}(x) \approx \frac{1}{2x\sqrt{\pi}} \exp(-x^2), \; x \gg 1 \tag{2.35}$$

(2.31) takes the form

$$P_{eASK,2} \simeq 0.5 \; \exp\left(-\frac{S^2}{8\sigma^2}\right) \tag{2.36}$$

Equation (2.36) is very suitable for comparison with corresponding expressions obtained for other modulation/detection schemes.

2.4.3 Coherent Detection of Optical ASK Signals

The coherent detection of ASK signals means, in general, that it is possible to restore the carrier at the receiver end, which will be synchronous and synphase with the carrier wave of the ASK signal. The coherent detection is based on the facts that the filtrated noise always consists of two orthogonal components and that the detection extracts the signal and the synphase noise component from the additive process, while the noise component that is in quadrature is being rejected.

The main difference between a coherent detection scheme and a noncoherent detection (envelope detection) scheme originates from the fact that the signal, passing through an IF filter, is not led to the envelope detector, but into a product modulator, where it is multiplied with the signal from a local electrical oscillator. The frequency of the local electrical oscillator signal is equal to the central frequency, ω_c, of the IF filter. The bandwidth of the IF filter is the same as that in noncoherent detection. After the multiplication process, the signal is passed through the baseband filter, and then it is sampled and compared with the threshold, as illustrated in Figure 2.3.

When the "1" state was emitted in the optical transmitter, the output from the product modulator, in accordance with (2.23), is a random process:

$$f(t) = f_s(t)\cos \omega_c t = Z \left[h(t) + x(t)\right]\cos \omega_c^2 \, t - Z\sin\omega_c t \, \cos \, \omega_c t \tag{2.37}$$

All components with frequencies higher than ω_c are rejected at the output of the baseband filter, so the decision is made for the next random process:

$$v_1(t) = Z[h(t) + x(t)] \tag{2.38}$$

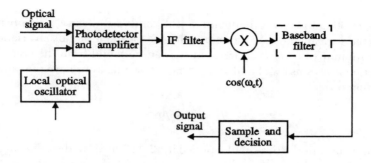

Figure 2.3 General scheme of optical receiver for coherent detection of ASK signals.

This process has the normal distribution with the mean value $\overline{v_1(t)} = Zh(t)$.

When the "0" state is sent, the following random process occurs after the receiver filter:

$$v_0(t) = Z\,x(t) \tag{2.39}$$

This process has the normal distribution, as well, with the mean value equal to zero. The sampling after the receiver filter is made at the moments t_0 corresponding to the half-widths of digit intervals. The sampled signal values are

$$h(t_0) = S_h \tag{2.40}$$

where the symbol interference is neglected.

The same logic as in the noncoherent detection case can be applied in the decision process, so the optimal threshold level can be determined by the method illustrated in Figure 2.2, but now the distribution functions corresponding to envelopes $v_1(t)$ and $v_0(t)$ are the same (Rayleigh distribution). To find the error probability in the coherent detection of ASK signals, the results, which refer to IM/DD lightwave systems (Appendix E), can be directly applied. With a common assumption that the numbers of the "0" and "1" states in the long pulse train are mutually equal, the expression for error probability can be written in the form

$$P_{eASK,2} = 0.5\left[1 - 0.5\Phi\left(\frac{S_h - b}{\sqrt{2}\sigma_{h1}}\right) - 0.5\Phi\left(\frac{b}{\sqrt{2}\sigma_{h0}}\right)\right] \tag{2.41}$$

where Φ is the Gaussian error function. Equation (2.41) can be simplified if we assume that $\sigma_{h1}^2 = \sigma_{h0}^2$. This is a reasonable assumption, because the influence of the local optical oscillator power is dominant in (2.8) and (2.9); hence, the optical power of the detected logical state has a minor influence on the total noise variance.

The optimum value of the threshold b_{opt} can be found from the cross-point of the distribution functions $w_0(v)$ and $w_1(v)$. Since these functions have the same shape, it is easily found that

$$b_{opt} = Z \, S_h/2 \qquad (2.42)$$

This optimum value of the threshold, applied in (2.41), leads to the following expression for error probability in the coherent detection of ASK optical signals

$$P_{eASK,2} = 0.5\left[1 - \Phi\left(\frac{S_h}{2\sqrt{2}\ \sigma_h}\right)\right] = 0.5 \ \mathrm{erfc}\left(\frac{S_h}{2\sqrt{2}\ \sigma_h}\right) \qquad (2.43)$$

It can be concluded from (2.41) that both terms on the right side give approximately equal contributions to the error probability, which is the characteristic feature of ASK coherent detection.

2.4.4 Error Probability in Optimal Detection of ASK Signals

To find the minimal error probability in ASK/heterodyne detection scheme, we can apply the general expression for minimal error probability in digital telecommunications system (see Appendix H). That expression has the form

$$P_{eMIN,2} = 0.5 \ \mathrm{erfc}\left\{\left[\frac{\overline{E}(1 - \kappa)}{4\nu_0}\right]^{1/2}\right\} \qquad (2.44)$$

where $\overline{E}$ is the averaged energy of both logical states, κ is the so-called correlation coefficient, and ν_0 is the power spectral density of the noise. The time dependence of the signal is expressed indirectly through the coefficient κ and energy $\overline{E}$. The minimal error probability, according to (2.44), can be reached only for the known time dependencies of both binary states and for the known moment of their appearance. That implicitly means that the signal phase in the optical receiver must be known in advance; hence, the optimal detection must be the coherent one.

Since the elementary states at the input of an optimal ASK receiver are given as

$$s_0(t) = 0 \ \text{or} \ s_1(t) = h(t) \ \cos \ \omega_0 t, \ 0 \le t \le T_d \qquad (2.45)$$

the averaged energy of both states is

$$\overline{E} = 0.5 \int_0^{T_d} [s_0^2(t) + s_1^2(t)]dt = 0.5\, E_1 \qquad (2.46)$$

where E_1 is the energy of the $s_1(t)$ elementary state. The level of likelihood function is chosen to be unity, so the optimum value of the decision threshold is

$$b_{opt} = 0.5 \int_0^{T_d} s_1^2(t)dt = 0.5\, E_1 \qquad (2.47)$$

The correlation coefficient is determined as

$$\kappa = \frac{1}{\overline{E}} \int_0^{T_d} s_0(t)s_1(t)dt = 0 \qquad (2.48)$$

Now, by introducing the values of $\overline{E}$ and κ in (2.44), the final expression for minimal bit error probability takes the form

$$P_{eMIN,2} = 0.5\ \mathrm{erfc}\left[\left(\frac{E_1}{8\nu_0}\right)^{1/2}\right] \qquad (2.49)$$

By comparing (2.49) with (2.43), we can see that the corresponding error probabilities will be identical if

$$\frac{S_h^2}{\sigma_h^2} = \frac{E_1}{\nu_0} \qquad (2.50)$$

Equation (2.50) can be realized in practice only if the receiver filter is matched to the elementary signal $s_1(t)$; for an unadapted filter the left side of (2.50) will always be lower than the right side. Hence, the error probability in an optimal receiver will always be lower than the error probability in a conventional receiver. The optimum coherent receiver offers another possibility for decreased error probability without increased signal power, which is by an increase in the elementary pulse duration. That way, the average energy of the signal is increased, which cannot be performed in a conventional coherent receiver.

2.4.5 Influence of Phase Noise on Detection of ASK Signals in Coherent Optical Receivers

The analysis in this section was made under the assumption that the local optical oscillator has the frequency and phase adjusted with the frequency and phase of the incoming ASK optical signal. However, the frequency and phase of the local optical oscillator deviate from some nominal values in a real coherent lightwave system. That is caused by the influence of the phase noise of the light source and by the variations of amplitude-phase characteristics of the transmission system in a hole. The phase shift between the nominal value, necessary for an ideal coherent detection, and a real value of local optical oscillator phase increases the error probability in a coherent detection scheme of any modulated digital signal [5, 6].

To make the quantitative analysis of such an effect in the heterodyne detection scheme of ASK signals we will start from (2.7) and (2.8). The phase difference can be expressed in the form

$$\Delta\varphi(t) = \varphi(t) + \vartheta_n(t) \tag{2.51}$$

where $\varphi(t)$ denotes the phase signal, and $\vartheta_n(t)$ presents the phase-shift fluctuations caused by the noises in the optical transmitter and the local optical oscillator. These phase fluctuations led to the spectral spreading, δf_ϑ, of the IF signal, even in cases without modulation.

In the noncoherent detection scheme of ASK signals, the extraction of the sent information is made by the envelope detector consisting of the ac/dc converter and the low-pass filter for elimination of the carrier frequency component. The output from such an envelope detector is relatively insensitive to phase fluctuations, $\vartheta_n(t)$. However, if the bandwidth, B, of the IF filter is comparable with spectral spreading δf_ϑ, the phase fluctuations will be transformed into amplitude ones, which is rather unfavorable for receiver sensitivity. To prevent that situation, it is necessary that the bandwidth be much higher than the frequency spreading.

In a coherent detection scheme, phase fluctuations lead to a serious decrease in optical receiver sensitivity. It is hard to generate referent electrical signal that is precisely synchronized with the IF signal; hence, the phase error, $\vartheta_n(t)$, between the IF signal and the referent one will be preserved. Thus, (2.38) and (2.39), which denote the amplitudes of the process before its sampling, become

$$v_1(t) = Z\left[h(t)\, \cos\, \vartheta_n(t) + x(t)\right] \tag{2.52}$$

and

$$v_0(t) = Z\, x(t) \tag{2.53}$$

The phase error, $\vartheta_n(t)$, can be considered as a noise that submits to the Gaussian distribution, with the mean value and variance equal to, respectively,

$$\overline{\vartheta_n(t)} = 0 \tag{2.54}$$

and

$$\overline{\vartheta_n^2(t)} = \sigma_n^2 \tag{2.55}$$

The probability density function of the considered phase noise is

$$w(\vartheta_n) = \frac{1}{\sqrt{2\pi}\,\sigma_n}\,\exp\!\left(-\frac{\vartheta_n^2}{2\sigma_n^2}\right) \tag{2.56}$$

Now we can evaluate the error probability by averaging (2.43) over the assemblage of phases ϑ_n, in accordance with (2.56). In such a way it is obtained that

$$P_{E,ASK} = \int_{-\pi}^{\pi} P_{gASK,2}(\vartheta_n) w(\vartheta_n)\, d\vartheta_n \tag{2.57}$$

or

$$P_{E,ASK} = \int_{-\pi}^{\pi} \frac{1}{\sqrt{2\pi}\sigma_n}\,\mathrm{erfc}\!\left(\frac{S_h \cos \vartheta_n}{2\sqrt{2}\,\sigma_h}\right)\,\exp\!\left(-\frac{\vartheta_n^2}{2\sigma_n^2}\right) d\vartheta_n \tag{2.58}$$

Equation (2.58) is related to the detection with an optimum threshold value. The integral (2.58) must be solved numerically, to obtain the parametric dependence of the error probability on the SNR, for different values of phase-noise variance (or phase noise standard deviation). As an illustration, four of these curves, for $\sigma_n = 0$, $\sigma_n = 0.15$, $\sigma_n = 0.2$ and $\sigma_n = 0.3$, are shown in Figure 2.4.

2.5 DIGITAL FREQUENCY MODULATION OF OPTICAL SIGNALS

2.5.1 Modulation of Optical Signals

The binary frequency modulation scheme produces FSK signals, with only two constant values, ω_1 and ω_2, of carrier frequency, which correspond to the logical states "1" and "0", respectively. The FSK signal can be obtained by the superposition of two ASK signals with different carrier frequencies. Hence, the detection of FSK signals can be successfully made by two separated receivers for the coherent or noncoherent detection of ASK signals.

There are two possibilities for frequency-shift keying, the "hard"-shift keying and the "soft" one. In the hard-shift keying scheme, the initial phases of two oscillators, which generate the corresponding modulating frequencies, can be in an arbitrary relationship, so the transitions from one frequency to another can be accompanied

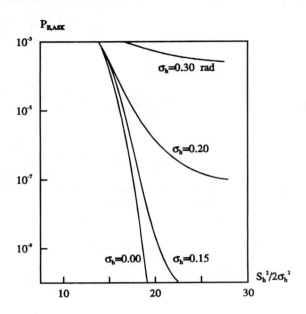

Figure 2.4 The dependence of error probability on signal-to-noise ratio for coherent ASK detection scheme.

by considerable amplitude jumps, an unfavorable effect due to the appearance of new spectral components [1]. In the soft-shift keying scheme, the obtained waveform of an FSK signal does not contain the amplitude jumps, but the frequency transitions are accompanied by the continual changes of the FSK signal amplitude. Such a case of FSK modulation, which is the most common in practice, is illustrated in Figure 2.5. This scheme is known as continuous phase FSK (CPFSK).

The FSK signal from Figure 2.5 has the mathematical form

$$s_{FSK}(t) = \begin{cases} S_P \cos \omega_1 t, \text{ for ``1'' state} \\ S_P \cos \omega_2 t, \text{ for ``0'' state} \end{cases}$$

(2.59)

and refers to the duration of the digit interval, T_d. The signal in (2.59) should be passed through a transmitter filter, where its spectrum would be narrowed. The main part of the FSK signal power is placed around the frequency $2/T_d$, and the equivalent transfer function, which is the product of the transfer functions of the transmitter filter, the optical fiber line, and the receiver filter, must have the bandwidth [1, 7]

$$B_t \geq 4/T_d$$

(2.60)

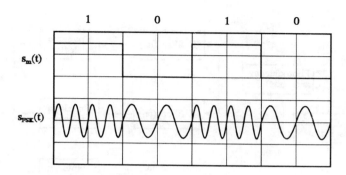

Figure 2.5 Waveform of an FSK signal.

This condition is essential for high-quality transmission, and it must be satisfied in coherent lightwave systems to prevent distortion of the transmitted signal. Thus, it is assumed that the influence of the dispersion effect in a single-mode optical fiber is small enough. It is also assumed that nonlinear effects in the optical fiber are negligible and do not influence the transmission quality. Thus, the only cause of the error during the decision process in a coherent optical receiver will be the additive noise coming to the input of the receiver filter.

2.5.2 Noncoherent Detection of FSK Signals

The general scheme of the optical receiver for noncoherent detection of binary FSK signals is shown in Figure 2.6. The receiver contains two parallel ports, with corresponding IF filters and the envelope detectors in each port. The difference signal from these two ports is coming to the sampler and the decision circuit. It is assumed that there is the exact information in receiver about digits interval and that the sampling is made at the moments that correspond to the middles of digits intervals. The filter in the upper port in Figure 2.6 has the central frequency $\omega_{c1} = \omega_1$ and the bandwidth B, while the filter in the lower port has the same bandwidth and the central frequency $\omega_{c2} = \omega_2$. Hence, the upper port filter allows passing only of the signal that corresponds to the "1" state, while the lower port filter allows passing only of the signal that corresponds to the "0" state.

If the logical "1" is sent in the considered digit interval $0 \leq t \leq T_d$, then we have the following additive process at the output of the upper port filter:

$$f_1(t) = h(t) \cos \omega_{c1}t + n_1(t) \tag{2.61}$$

where $n_1(t)$ is filtrated additive noise with variance σ_1^2. At the same time, only the filtrated noise

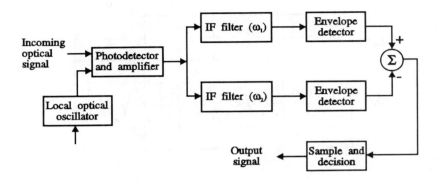

Figure 2.6 General scheme of FSK receiver for noncoherent detection.

$$f_2(t) = n_2(t) \tag{2.62}$$

appears at the output of the filter with central frequency ω_2. The variance of this noise will be σ_2^2. At the moment of sampling (for $t = t_0$), the process $f_1(t)$, which represents the sum of the narrowband signal and the narrowband noise, will have the Rice distribution (see Appendix D)

$$w_1(r_1) = \frac{r_1}{\sigma_1^2} \exp\left(-\frac{r_1^2 + S^2}{2\sigma_1^2}\right) I_0\left(\frac{r_1 S}{\sigma_1^2}\right) \tag{2.63}$$

where r_1 is the envelope of the process at the output of the filter in the upper port, while S is the signal amplitude at the sampling moment, that is, $S = h(t_0)$. At the same time, the envelopes of the process $f_2(t)$ will have the Rayleigh distribution (see Appendix C), so we have

$$w_2(r_2) = \frac{r_2}{\sigma_2^2} \exp\left(-\frac{r_2^2}{2\sigma_2^2}\right) \tag{2.64}$$

where r_2 is the envelope of the process at the output of the lower port filter. Since the bandwidths of the filters do not overlap, the noises $n_1(t)$ and $n_2(t)$ are statistically independent. Hence, the envelopes $r_1(t)$ and $r_2(t)$ are statistically independent, as well.

Since just the difference of $r_1(t_0) - r_2(t_0)$ is sampled in an FSK receiver, and then the decision is made, the decision threshold is set to be zero ($b = 0$). Hence, the state 1 or 0 will be chosen depending on whether the difference is higher or lower than zero, respectively. To find the error probability in such a case of detection, the illustration given in Figure 2.7 can be used.

If $r_1(t_0) = r_1$ at the decision moment, the partial conditional probability, $P(0/1)$, of a false decision is equal to the shadowed area under the Rayleigh curve, and the following relation is valid:

$$P(0/1)\Big|_{r_1} = \int_{r1}^{\infty} w_2(r_2)dr_2 = \exp\left(-\frac{r_1^2}{2\sigma_2^2}\right) \qquad (2.65)$$

Since the instantaneous value of envelope $r_1(t)$ can take any positive value, the total conditional probability, $P(0/1)$, can be found by the summation over all the region of variation of variable r_1, so it can be written that

$$P(0/1) = \int_0^{\infty} w_1(r_1)\left[\int_{r1}^{\infty} w_2(r_2)dr_2\right]dr_1 = \int_0^{\infty} w_1(r_1)\exp\left(-\frac{r_1^2}{2\sigma_2^2}\right)dr_1 \qquad (2.66)$$

The solution of the integral in (2.66) can be found by using the substitution $r_1 = z/\sqrt{2}$; thus, (2.66) becomes

$$P(0/1) = 0.5\int_0^{\infty} \frac{z}{\sigma_1^2} \exp\left(-\frac{z^2}{2\sigma_2^2}\right) \exp\left(-\frac{S^2}{2\sigma_1^2}\right) I_0\left(\frac{zS}{\sqrt{2}\sigma_1^2}\right) dz \qquad (2.67)$$

The square term from (2.67) can be eliminated by using the substitution $S_1 = S/\sqrt{2}$. Now, with the earlier assumption that $\sigma_1 = \sigma_2$, the function under the

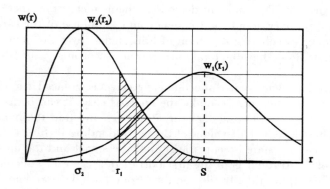

Figure 2.7 Probability density functions in noncoherent detection of FSK signals.

integral takes the form of a Rice distribution. Since the integral of the Rice function over all positive values is equal to unity, (2.67) takes the final form

$$P(0/1) = 0.5 \, \exp\left(-\frac{S^2}{4\sigma^2}\right) \tag{2.68}$$

The same expression is obtained for conditional probability $P(1/0)$, under the assumption that only the noise is passing through the upper port filter of the receiver, while the sum of the signal and the noise is passing through the lower port filter. Thus, the expression for error probability in a noncoherent detection scheme of FSK signals can be written in the form

$$P_{eFSK,2} = [P(1) + P(0)] \, P(0/1) = 0.5 \, \exp\left(-\frac{S^2}{4\sigma^2}\right) \tag{2.69}$$

The obtained expression for error probability can be compared with (2.36), which refers to the noncoherent detection of ASK signals, since the operational conditions are identical in both cases (the same filter bandwidths and the powers of filtrated noise). It can be seen that ASK signal detection requires signal power two times higher than the signal power in the FSK detection scheme, in order to achieve an equal error probability. Hence, the receiver sensitivity of an FSK receiver with noncoherent detection is 3 dB higher than the receiver sensitivity of an ASK receiver with noncoherent detection.

2.5.3 Coherent Detection of FSK Optical Signals

The general scheme of the optical receiver for coherent detection of FSK signals is given in Figure 2.8. The coherent detection scheme means that there is information in the optical receiver about the frequency and the phase of the carrier wave. The main advantage of coherent detection of FSK signals over the noncoherent scheme is the rejection of one-half the total noise power. Accordingly, the SNR will increase for 3 dB.

The envelope detector in a noncoherent receiver is replaced in a coherent detector by the product modulator. To make the analysis easy, it can be assumed that the local carriers $\cos\omega_{c1}$ and $\cos\omega_{c2}$ are the ideal ones and that the bandwidths of the baseband filters are the half-values of the corresponding IF filters.

When the "1" state is sent, the sum of the signal and the noise is present at the output of the IF filter is in the first port of the receiver, while only the noise is present at the output of the IF filter in the second port. The random processes at the outputs of product modulators, belonging to the first and the second ports, can be expressed, respectively, as

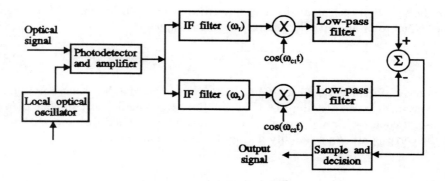

Figure 2.8 The general scheme of optical receiver for coherent detection of FSK signals.

$$f_1(t)\cos \omega_{c1}t = [h(t) + x_1(t)]\cos^2 \omega_{c1}t - y_1(y)\sin \omega_{c1}t \cos \omega_{c1}t \qquad (2.70)$$

$$f_2(t)\cos \omega_{c2}t = x_2(t)\cos^2 \omega_{c2}t - y_2(t)\sin \omega_{c2}t \cos \omega_{c2}t \qquad (2.71)$$

After the multiplication the low-pass filtration is made, all components with higher frequencies are rejected. Thus, the baseband random processes $v_1(t)$ and $v_2(t)$ are obtained at the outputs of the baseband filters. These processes have the Gaussian distributions of instantaneous values. If the value of transimpedance Z is taken to be unity, the processes at the filters outputs, when the "1" state was sent, have the forms

$$v_1(t) = h(t) + x_1(t) \qquad (2.72a)$$
$$\text{for "1"}$$
$$v_2(t) = x_2(t) \qquad (2.72b)$$

while if the "0" state was sent, we have

$$v_1(t) = x_1(t) \qquad (2.73a)$$
$$\text{for "0"}$$
$$v_2(t) = h(t) + x_2(t) \qquad (2.73b)$$

The difference $v_1(t) - v_2(t)$ is sampled at the moment $t = t_0$, and the following values are obtained:

$$v_1(t_0) - v_2(t_0) = S_h + (x_1 - x_2), \, a_n = 1 \qquad (2.74a)$$
$$v_1(t_0) - v_2(t_0) = -S_h + (x_1 - x_2), \, a_n = 0 \qquad (2.74b)$$

It can be seen that the sampled values have the normal distributions with the mean values equal to S_h and $-S_h$. The corresponding variance is

$$\overline{(x_1 - x_2)^2} = \overline{x_1^2} - \overline{2x_1x_2} + \overline{x_2^2} = \overline{x_1^2} + \overline{x_2^2} = \sigma_{h1}^2 + \sigma_{h0}^2 \qquad (2.75)$$

If the power of the local optical oscillator is high enough, the difference between the noise variances corresponding to "1" and "0" is very small (even smaller than in the case of coherent ASK detection). Hence, it can be assumed that $\sigma_{h1} \simeq \sigma_{h0} = \sigma_h$, and the corresponding probability density functions take the forms

$$w_1(x) = \frac{1}{\sqrt{4\pi\sigma_h^2}} \exp\left(-\frac{(x - S_h)^2}{4\sigma_h^2}\right), \; a_n = 1 \qquad (2.76)$$

$$w_2(x) = \frac{1}{\sqrt{4\pi\sigma_h^2}} \exp\left(-\frac{(x + S_h)^2}{4\sigma_h^2}\right), \; a_n = 0 \qquad (2.77)$$

The functions in (2.76) and (2.77) are shown in Figure 2.9. The shadowed area to the left of point $x = 0$ is equal to the conditional probability $P(0/1)$. This probability is determined as

$$P(0/1) = \int_{-\infty}^{0} w_1(x)dx = 0.5 \; \text{erfc}\left(\frac{S_h}{2\sigma_h}\right) \qquad (2.78)$$

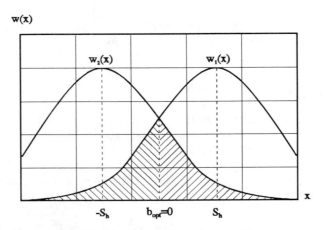

Figure 2.9 Probability density functions for coherent detection FSK signals and the optimal threshold value.

Since the ports in the coherent FSK receiver are symmetrical, the conditional probability $P(1/0)$ is equal to the value $P(0/1)$, so the total error probability for the coherent detection scheme of FSK signals is determined as

$$P_{eFSK,2} = [P(1) + P(0)] \, P(0/1) = 0.5 \, \text{erfc}\left(\frac{S_h}{2\sigma_h}\right) \tag{2.79}$$

Now, by comparing (2.79) with the equivalent expression for the error probability of coherent detection of ASK signals, it can be seen that the receiver sensitivity of a coherent FSK receiver is 3 dB higher than the receiver sensitivity of a coherent ASK receiver. This relationship is valid for peak values of optical power. But the situation is quite different if the average values are taken instead of peak ones. In such a case, the average value of an FSK signal is twice the average value of an ASK signal. Hence, the equal average values of the optical powers cause the equal error probabilities. But the optimum decision level is constant for coherent FSK detection, and it is the main advantage of the FSK coherent detection scheme over the ASK coherent detection scheme (the decision threshold is not constant in the ASK coherent detection scheme).

2.5.4 Error Probability in Optimal Detection of FSK Signals

To find the error probability in the optimum FSK optical receiver, we will again start from (2.44). But now there are two nonzero elementary states, $s_1(t)$ and $s_2(t)$. That means [1] the optimum receiver must have two parallel ports with two correlators or two matched filters. The averaged energies of both elementary states are identical, so we have

$$\overline{E} = 0.5\int_0^{T_d} [s_1^2(t) + s_2^2(t)]dt = 0.5(E_1 + E_2) = E_1 \tag{2.80}$$

The correlation coefficient κ depends on the choice of frequencies ω_1 and ω_2. In the case when

$$\omega_1 - \omega_2 = k\pi/T_d, \; \omega_1 + \omega_2 = m\pi/T_d; \; k, m = 1, 2, \ldots \tag{2.81}$$

the correlation coefficient κ becomes

$$\kappa = \frac{1}{E}\int_0^{T_d} s_1(t)s_2(t)dt = 0 \tag{2.82}$$

In accordance with (2.44), the error probability in the optimum FSK optical receiver takes the form

$$P_{eMIN,2} = 0.5 \ \mathrm{erfc}\left[\left(\frac{\overline{E}}{4\nu_0}\right)^{1/2}\right] = 0.5 \ \mathrm{erfc}\left[\left(\frac{\overline{E_1}}{4\nu_0}\right)^{1/2}\right] \tag{2.83}$$

We can see from (2.49) and (2.83) that error probabilities for optimum detection of ASK and FSK signals are equal if the averaged signal energies are equal. If the energies of the elementary states are taken to be equal instead of the equality of average energies, however, the energy of the elementary signal in the ASK system must be twice the energy of the elementary signal in FSK system for the same error probabilities. Hence, as we expected, the sensitivity of the optimum FSK receiver is 3 dB higher than the sensitivity of the optimum ASK receiver [8, 9].

2.5.5 Influence of Phase Noise on Detection of FSK Signals in a Coherent Optical Receiver

It has already been mentioned that the frequency and the phase of a local optical oscillator in a real coherent lightwave system deviates from their nominal values due to influence of the phase noise of the light source and distortions of amplitude-phase characteristics of the lightwave system in a hole. Every deviation of the phase from its nominal value (referring to an ideal coherent detection) increases the error probability in an FSK system [10–12].

The influence of the phase noise on the error probability in an FSK system can be considered by using the same method as for an ASK system. Thus, (2.51), (2.52), and (2.56) should be taken into account. The amplitudes of the random processes before the sampling can be obtained from (2.73) and (2.74) and have the form

$$v_1(t) = Z[h(t)\cos\ \vartheta_n(t) + x(t)], \ a_n = 1 \tag{2.84}$$

$$v_2(t) = -Z[h(t)\cos\ \vartheta_n(t) + x(t)], \ a_n = 0 \tag{2.85}$$

The error probability in the presence of phase noise in an FSK heterodyne detection scheme can be found by averaging (2.79) over all the values of ϑ_n, according to the distribution function in (2.56). In such a way it is obtained that

$$P_{E,FSK} = \int_{-\pi}^{\pi} P_{eFSK,2}(\vartheta_n)w(\vartheta_n)d\vartheta_n \tag{2.86}$$

or

$$P_{E,FSK} = \int_{-\pi}^{\pi} \frac{1}{2\sqrt{2\pi}\sigma_n} \, \text{erfc}\left(\frac{S_h \cos\vartheta_n}{2\sigma_h}\right) \exp\left(-\frac{\vartheta_n^2}{2\sigma_n^2}\right) d\vartheta_n \qquad (2.87)$$

The parametric dependence of error probability on the signal-to-noise powers ratio $(S_h^2/2\sigma_h^2)$ for four different values of phase noise standard deviations ($\sigma_n = 0$, 0.25, 0.4, and 0.5 rad) are shown in Figure 2.10. The curves in Figure 2.10 were obtained by numerical integration of (2.87). It can be easily concluded that the phase-noise influence on the characteristics of an FSK optical receiver is less than its influence on the characteristics of an ASK optical receiver, so the FSK receiver is more robust than the ASK optical receiver.

2.6 DIGITAL PHASE MODULATION OF OPTICAL SIGNALS

2.6.1 Modulation of Optical Signals

The digital phase modulation, or phase-shift keying (PSK), can be treated as a form of conventional phase modulation. The PSK signal, $s_{PSK}(t)$, can be defined as a real part of the complex phase-modulated signal, or

$$s_{PSK}(t) = \text{Re}\left\{\sum_{-\infty}^{\infty} h_p(t - nT_d) \, \exp[\,j(\omega_0 t + \varphi_n)]\right\} \qquad (2.88)$$

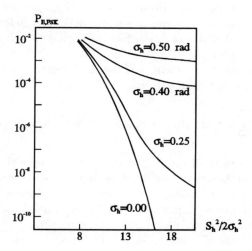

Figure 2.10 Error probability for FSK detection of optical signals.

where $h_p(t)$ is a rectangular pulse with amplitude S and duration T_d, ω_0 is the frequency of the carrier wave, and (φ_n) represents the sequence of initial phases. The sequence of initial phases presents an information content, where the individual elements take the discrete values within the interval $[1 - 2\pi]$. The real part of (2.88) is

$$s_{PSK}(t) = \cos\ \omega_0 t \sum_{-\infty}^{\infty} h_p(t - nT_d)\cos\ \varphi_n + \sin\ \omega_0 t \sum_{-\infty}^{\infty} h_p(t - nT_d)\sin\ \varphi_n \qquad (2.89)$$

If we take that $\cos\varphi_n$ is equal to the element a_n of the information content of the baseband digital signal, and if $\sin\varphi_n$ is equal to element b_n of the information content of another baseband signal, then (2.89) can be considered as the signal with quadruple amplitude modulation (QAM). In the binary phase keying scheme of the signal represented by (2.89), the initial phases take only two values: 0 or π. Thus, it can be written that

$$s_{PSK}(t) = \sin\ \omega_0 t \sum_n a_n h_p(t - nT_d) \qquad (2.90)$$

where the element of information content can take only two values, or

$$a_n = \begin{cases} \cos\ 0 = 1 \\ \cos\ \pi = -1 \end{cases} \qquad (2.91)$$

Hence, the binary PSK signal presents a simple amplitude-modulated signal with suppressed carrier wave. The spectral width of such a defined PSK signal is equal to twice the spectral width of a modulating signal, which consists of the train of pulses with the shape $h_p(t)$.

The expression for quaternary phase modulation can be obtained in a similar way. If φ_n takes the values $\varphi_n \epsilon(0, \pi/2, -\pi/2, \pi)$, (2.89) becomes

$$s_{4PSK}(t) = \cos\ \omega_0 t \sum_{-\infty}^{\infty} a_n h_p(t - nT_d)\ \cos\ \varphi_n + \sin\ \omega_0 t \sum_{-\infty}^{\infty} b_n h_p(t - nT_d)\ \sin\ \varphi_n \qquad (2.92)$$

The elements a_n and b_n of the information content take the values $a_n \epsilon(1, -1)$ and $b_n(1, -1)$. The total bandwidth of a 4-PSK signal is still equal to double the spectral width of a modulating signal, which consists of the train of pulses $h_p(t)$. On the other hand, the frequency bandwidths of 2-PSK and 4-PSK signals are equal to the bandwidth of an ASK signal.

The application of a quaternary modulation scheme instead of a binary one brings the benefit of the frequency bandwidth for the same bit rates. Further benefit can be achieved by an increase in the number of elementary initial phases to 8 or

16, for example, with unchanged digit duration T_d. In such a case, 8-valued PSK and 16-valued PSK signals are obtained. The increase in the number of possible phase states does not increase the frequency bandwidth of the modulated signal but does increase the error probability in the system, which we will discuss later.

2.6.2 Coherent Detection of Optical Binary PSK Signals

A coherent FSK optical receiver contains two parallel ports, two IF filters, and two local electrical oscillators. The doubling of these elements is caused by the fact that there are two different frequencies corresponding to the elementary binary states in a modulating signal. But if there is only one frequency, and information is contained in the changes of the initial phase of the carrier optical wave, only one receiver port can perform all the necessary functions. This reduction leads to simplification of the receiver and to reduction of the additive noise power before the sampler and decision circuit, because there is no noise from another parallel port, as in an FSK receiver. Thus, it is clear that the detection of a binary PSK signal will bring a 3-dB benefit in receiver sensitivity, compared with detection of a binary FSK signal.

The general scheme of an optical receiver for coherent detection of PSK signals, or so-called CPSK optical receiver, is shown in Figure 2.11. After the optoelectronic conversion, the mixed signal passes through the same stages as in a coherent ASK receiver.

When the binary PSK signal, given by (2.89), is sent by the optical transmitter, the signal after the IF filter [which has the central frequency $\omega_c = \Delta\omega$, according to (2.8)] has the form

$$s(t) = h(t)\,\cos(\omega_c t + \varphi_n),\ \varphi_n\epsilon(0,\ \pi) \tag{2.93}$$

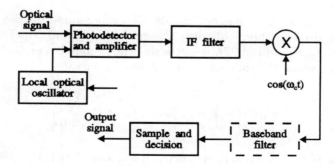

Figure 2.11 General scheme of binary CPSK optical receiver.

This signal is accompanied by the filtrated Gaussian noise, which has the variance σ_h^2. The signal power does not depend on the information content. At the output of the product modulator, there is the random process described by

$$f(t) = \{[a_n(t) + x(t)]\cos \omega_c t - y(t)\sin \omega_c t\}\cos \omega_c t \qquad (2.94)$$

where a_n takes either the value 1 or the value -1. After the operations of multiplication and filtration, the random process $v(t)$ is generated, which should be sampled at the moments corresponding to the middles of digits intervals. The amplitude of a sample at the moment t_0 is $|h(t_0| = ZS_h$, so we have

$$v(t_0) = \begin{cases} S_h + x(t_0), & a_n = 1 \\ -S_h + x(t_0), & a_n = -1 \end{cases} \qquad (2.95)$$

(It is assumed that transimpedance Z is equal to unity.)

When "1" was sent, the statistical mean value of the samples presented by (2.95) is S_h; if "0" was sent, that mean value is $-S_h$ (do not forget that "1" and "0" correspond to the values $a_n = 1$ and $a_n = -1$, respectively). The corresponding probability density functions for these two cases are given by

$$w_1(x) = \frac{1}{\sqrt{2\pi}\,\sigma} \exp\left(-\frac{(x - S_h)^2}{2\sigma^2}\right), \; a_n = 1 \qquad (2.96)$$

and

$$w_0(x) = \frac{1}{\sqrt{2\pi}\,\sigma} \exp\left(-\frac{(x + S_h)^2}{2\sigma^2}\right), \; a_n = -1 \qquad (2.97)$$

Figure 2.9 can be used to illustrate the decision process and the determination of the optimum decision threshold, since the curves $w_2(x)$ and $w_1(x)$ in Figure 2.9 correspond to (2.96) and (2.97), respectively. The conditional probability that "0" state is chosen when "1" state was sent is determined as

$$P(0/1) = \int_{-\infty}^{0} w_1(x)dx = 0.5 \; \text{erfc}\left(\frac{S_h}{\sqrt{2}\sigma_h}\right) \qquad (2.98)$$

Because of the condition of symmetry, it is always $P(0/1) = P(1/0)$; hence, the equivalent content of the "1" and "0" states in the signal leads to the next expression for error probability in a binary PSK system:

$$P_{ePSK,2} = [P(0) + P(1)]P(0/1) = 0.5 \; \text{erfc}\left(\frac{S_h}{\sqrt{2}\sigma_h}\right) \qquad (2.99)$$

This is the expected result, which confirms the 3-dB improvement in receiver sensitivity in comparison with the coherent FSK receiver. At the same time, the frequency bandwidth of the PSK signal is half the bandwidth of the FSK signal.

When the ratio S_h/σ_h is relatively high, (2.99) can be replaced by the approximation

$$P_{ePSK,2} = \frac{\sigma_h}{S_h\sqrt{2\pi}} \; \exp\left(-\frac{S_h^2}{2\sigma_h^2}\right) \qquad (2.100)$$

which is very suitable for comparison with the expression for error probability in a multivalued (M-valued) PSK detection scheme.

2.6.3 Coherent Detection of Optical M-Valued PSK Signals

The equivalent transmission rate can be increased by an increase in the initial discrete phase states, while the frequency bandwidth of the signal stays unchanged. However, an increased number of phase states causes an increased error probability for the given SNR. Hence, there is a question of what is the upper limit of the transmission capacity increase, or, rather, what is the limit of tolerable deterioration of optical receiver sensitivity.

The general scheme of an M-valued PSK system (M > 2) is shown in Figure 2.12. It can be assumed that the elementary signal after the receiver IF filter has a rectangular shape with the amplitude S_h and that

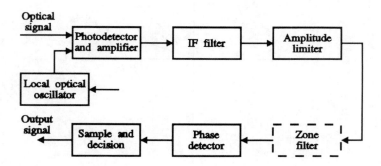

Figure 2.12 General scheme of coherent M-valued PSK optical receiver.

$$s(t) = S_h \cos(\omega_c t + \varphi_k),\ 0 \le t \le T_d \qquad (2.101)$$

$$\varphi_k \epsilon \left(0,\ \frac{2\pi}{M},\ \frac{4\pi}{M},\ \ldots,\ \frac{2(M-1)\pi}{M} \right) \qquad (2.102)$$

Such a signal together with the additive noise with variance σ^2 excites an ideal limiter. The limiter is followed by a so-called *zone filter*, which removes all the spectral components appearing in the process of amplitude limitation. The combination of the limiter and the zone filter removes all the amplitude variations of the filtrated process $f(t)$ but preserves its instantaneous frequency and instantaneous phase. The signal $f_\ell(t)$ from the limiter is led to the phase detector, where its instantaneous phase is measured. At the most suitable time-moment, this random phase is compared with one of the referent discrete phases, φ_k, and the decision about phase level is made.

The situation that can appear at the moment of decision is illustrated in Figure 2.13, where the four phasers are presented. These phasers are equidistant in the complex plane and correspond to the group of initial phases. Each signal component is followed by the additive noise (but is illustrated by only one noise phaser). The noise consists of two components, one in phase with the signal and the other in quadrature with the signal. These noise components contribute both to the signal amplitude distortion and to the signal phase distortion. The optimum decision thresholds, corresponding to relevant values of the initial phase, are denoted by the dashed lines in Figure 2.13. The space between the two dashed lines corresponds to the region of the maximum a posteriori probability, which refers to the value of the measured phase from that region [1].

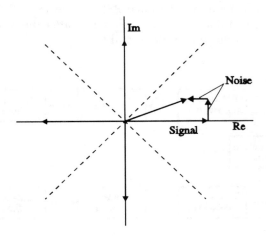

Figure 2.13 The phaser diagram corresponding to detection of 4-PSK signals.

If the transmitted signal has the initial phase φ_k, it can be assumed that the initial phase of the noise is φ_k as well, because the noise can have an arbitrary initial phase. The filtrated noise can be expressed in the form

$$n(t) = x(t)\cos(\omega_c t + \varphi_k) - y(t)\sin(\omega_c t + \varphi_k) \tag{2.103}$$

The random process after the IF filter has the form

$$f(t) = s(t) + n(t) = r(t)\cos[\omega_c t + \varphi_k + \eta(t)] \tag{2.104}$$

The instantaneous phase of a random process consists of a deterministic component, φ_k, and a random component,

$$\eta(t) = \arctan\frac{y(t)}{h(t) + x(t)} \tag{2.105}$$

The variations of the envelope $r(t)$ of the filtrated signal do not influence the measured value of the instantaneous phase, due to the existence of an ideal limiter before the phase detector. The measured value of the phase will not be equal to φ_k, because of noise influence, but will fluctuate around that value, so we have

$$\varphi = \varphi(t_0) = \varphi_k + \arctan\frac{y(t_0)}{S_h + x(t_0)} \tag{2.106}$$

To make the correct detection, it is necessary that the measured phase φ is within the region of the maximum a posteriori probability, that is,

$$\left(\varphi_k - \frac{\pi}{M}\right) < \varphi < \left(\varphi_k + \frac{\pi}{M}\right) \tag{2.107}$$

All regions in the phase plane from Figure 2.13 have the equivalent areas; thus, the error probability per digit can be evaluated as

$$P_{ePSK,M} = P\left(|\varphi| > \frac{\pi}{M}\right) = 1 - \int_{-\pi/M}^{\pi/M} w(\varphi)d\varphi \tag{2.108}$$

The probability density function, $w(\varphi)$, refers to the case considered in Appendix D [see (D.16)]. For higher SNRs and for relatively small deviations of the instantaneous phase from its mean value, (D.16) can be simplified. Thus, the following simplified form is obtained:

$$w(\varphi) = \frac{1}{\sqrt{2\pi}} \frac{S_h \cos \varphi}{\sigma_h} \exp\left(-\frac{S_h^2 \sin^2 \varphi}{2\sigma_h^2}\right), \; |\varphi| < \pi/4 \qquad (2.109)$$

The integral in (2.108) can be solved by the substitution $z = S_h \sin\varphi/\sqrt{2\sigma_h}$, so we have

$$\int_{-\pi/M}^{\pi/M} w(\varphi)d\varphi = \mathrm{erf}\left(-\frac{S_h \sin \dfrac{\pi}{M}}{\sqrt{2}\sigma_h}\right) \qquad (2.110)$$

The error probability, $P_{ePSK,M}$, now becomes

$$P_{ePSK,M} = \mathrm{erfc}\left(-\frac{S_h \sin \dfrac{\pi}{M}}{\sqrt{2}\sigma_h}\right) \quad M \geq 4 \qquad (2.111)$$

In such a way an approximate relation for error probability for coherent detection of an M-level PSK signal is obtained. The approximation is better for higher ratios of peak values of signal and noise and for narrower region of a posteriori probability (or for a higher number of elementary states M).

The function erfc(z) in (2.111) can be approximated by an exponential function if the SNR is high enough. In such a case, the expression for error probability becomes

$$P_{ePSK,M} = \sqrt{2/\pi}\, \frac{\sigma_h}{S_h \sin \dfrac{\pi}{M}} \exp\left(-\frac{S_h^2 \sin^2 \dfrac{\pi}{M}}{2\sigma_h^2}\right) \qquad (2.112)$$

For the binary modulation scheme ($M = 2$), (2.112) takes the form

$$P_{ePSK,2} = \sqrt{2/\pi}\, \frac{\sigma_h}{S_h} \exp\left(-\frac{S_h^2}{2\sigma_h^2}\right) \qquad (2.113)$$

Equation (2..113) is rather close to (2.100), which is obtained for the detection of binary PSK signals. If $M = 4$ (quaternary PSK), (2.112) takes the form

$$P_{ePSK,4} = \sqrt{4/\pi}\, \frac{\sigma_h}{S_h} \exp\left(-\frac{S_h^2}{4\sigma^2}\right) \qquad (2.114)$$

Hence, the ratio of peak values of signal and noise should be doubled in order to reach the equivalent error probability per elementary signal as it was in the 2-PSK detection scheme. (The elementary 4-PSK signal carries double the quantity of information as the elementary 2-PSK signal.)

The error probability per elementary signal will be the same for all M-valued PSK systems if the term in the exponent of (2.112) is a constant. That can be reached by the contemporary increase of SNR for every increase of the number M. The factor of the SNR increase should be 2 for 4-PSK detection and 6.8 for 8-PSK detection, compared with the ratio for 2-PSK detection. The factor of increase is 26.5 and 102 for 16-PSK and 32-PSK detections, respectively [8]. Thus, the increase of the SNR has a nonlinear character, and the 32-PSK detection scheme is noneconomic from today's point of view.

2.6.4 Optimal Detection of Optical Binary PSK Signals

We will use (2.44) to find the error probability in an optimal PSK optical receiver. The receiver can expect one of two elementary states:

$$s_1(t) = h(t)\sin \omega_0 t, \ 0 \leq t \leq T_d$$

or   (2.115)

$$s_2(t) = -h(t)\sin \omega_0 t$$

Since the frequencies of the two signals in (2.115) are identical, the optimal PSK receiver can operate with two ports or with only one port. In either case, the decision threshold will be equal to zero. Because the elementary states have equal energies, the averaged energy of both states, $\overline{E}$, is equal to the energy, E_1, of one elementary state. The correlation coefficient, κ, is

$$\kappa = \frac{1}{E} \int s_1(t)s_2(t)dt = -1 \tag{2.116}$$

Insertion of the values for $\overline{E}$ and κ in (2.44) gives the following equation for error probability in an optimal PSK receiver:

$$P_{eMIN,2} = 0.5 \ \mathrm{erfc}\left[\left(\frac{\overline{E}}{2\nu_0}\right)^{1/2}\right] = 0.5 \ \mathrm{erfc}\left[\left(\frac{E_1}{2\nu_0}\right)^{1/2}\right] \tag{2.117}$$

This result is the expected one and means that error probability in a binary PSK system is lower than the error probability in any other digital modulation/detection scheme, for a given signal energy and a given spectral density of the noise. The increase in receiver sensitivity of an optimum PSK receiver is 3 dB compared with an optimum FSK receiver sensitivity. To illustrate the advantage of a PSK detection scheme over ASK and FSK detection schemes, we can say that the value of an erfc argument equal to 10 would cause an error probability of about 10^{-3} for an ASK or a FSK system, while the error probability for a PSK system would be below 10^{-5}.

2.6.5 Influence of Phase Noise on Detection of PSK Signals

Equations (2.51), (2.52), and (2.56) are essential for consideration of phase-noise influence and the error probability in PSK detection schemes. The PSK signals are very sensitive to the influence of the phase noise, because the entire information is contained in the signal phase changes [13]. Since the signal amplitude at the sampling moments depends on the value of phase error, $\vartheta_n(t)$, (2.95) takes the form

$$v(t_0) = \begin{cases} S_h \cos \vartheta_n(t_0) + x(t_0), & a_n = 1 \\ -S_h \cos \vartheta_n(t_0) + x(t_0), & a_n = -1 \end{cases} \tag{2.118}$$

Because the detection scheme is the same, the optimum decision threshold is equal to zero. The error probability can be found by averaging the instantaneous error probability, given by (2.99), over all possible values of $\vartheta_n(t)$. Keeping in mind that the phase error, $\vartheta_n(t)$, has a normal distribution, it can be obtained that

$$P_{E,PSK} = \int_{-\pi}^{\pi} 0.5 \, w(\vartheta_n) \text{erfc}\left(\frac{S_h \cos \vartheta_n}{\sqrt{2}\sigma_h}\right) d\vartheta_n \tag{2.119}$$

where $w(\vartheta_n)$ is given by (2.56). The integral in (2.119) can be solved numerically. The results of calculating four values of phase-noise standard deviations ($\sigma_n = 0.0$, 0.25, 0.4, and 0.5 rad) are shown in Figure 2.14. More precise results for error probability can be obtained by choosing a more appropriate expression for the probability density function, $w(\vartheta_n)$, such as (4.78), which takes into account the influence of the PLL on phase-noise characteristics.

By comparing the diagrams in Figure 2.10 and Figure 2.14, which are related to binary FSK and binary PSK receivers, respectively, it can be seen that the curves from Figure 2.14 can be obtained by a translation of curves from Figure 2.10 for 3 dB on the left. Hence, the influence of phase noise on the detection of FSK and PSK signals is identical, that is, the difference in receiver sensitivities stays the same (3 dB).

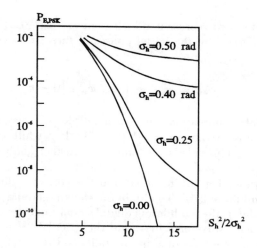

Figure 2.14 The influence of phase noise on the error probability for coherent detection of PSK signals.

2.7 DIFFERENTIAL DIGITAL PHASE MODULATION OF OPTICAL SIGNALS

2.7.1 Differential Digital Phase Modulation

The differential phase-shift keying (DPSK) scheme, accompanied by differentially coherent PSK detection (DCPSK), is introduced with the main aim of improving the characteristics of a PSK modulation/detection scheme and removing the difficulties related to the random deviations in the amplitude-phase characteristics of a transmission system.

The main difference between PSK and DPSK modulation schemes is in how information is represented. The elements of information content are represented by the absolute values of the initial phases in a PSK modulator; in a DPSK modulator, the transitions between neighboring elementary states correspond to the elements of information content. Hence, the absolute values of the initial phases no longer play an important role. In accordance with that, it is not necessary to measure the absolute phase of an incoming elementary signal in a CPSK detection scheme, but it is enough to compare this phase with the phase from the preceding digit interval [8].

We will briefly consider the DPSK modulation scheme. It can be assumed that the following signals are sent in digit intervals of n and $n - 1$:

$$s_{n-1}(t) = S_p \cos(\omega_0 t + \varphi_{n-1}), \quad (n - 1)T_d < t < nT_d \tag{2.120}$$

$$s_n(t) = S_p \cos(\omega_0 t + \varphi_n), \quad nT_d < t < (n + 1)T_d \tag{2.121}$$

The basic intention is to represent the element a_n $(a_n\epsilon[-1, 1])$ of the information content by the difference of the initial phases of these two neighboring elementary signals, or by

$$\Delta\varphi_n = \varphi_n - \varphi_{n-1}; \quad \varphi_n, \ \varphi_{n-1}\epsilon \ [0 - \pi] \qquad (2.122)$$

One of the possible modulation rules in a DPSK scheme can be that the value $\Delta\varphi_n = \pi$ corresponds to value $a_n = -1$ (logical "1"), while $\varphi_n = 0$ corresponds to $a_n = 1$ (logical "0"). This rule is convenient for further analysis, so we will assume that it is applied.

The waveforms of a DPSK signal and the corresponding modulating digital signal are given in Figure 2.15. To obtain the shown modulated signal at the output of the phase modulator, the modulating digital signal can be transformed into another "differentially coded" waveform first, and then the DPSK modulation rule should be applied. A new element, b_n, of a differentially coded signal presents the "module-two" sum of elements a_n and a_{n-1}.

An ordinary DPSK signal is generated by the acting of a differentially coded signal on the ideal phase modulator. This signal, shown in Figure 2.15, has the constant amplitude S_p and discrete phase changes. Such a signal propagates through the optical fiber line and comes to the optical receiver, where either a heterodyne or a homodyne detection scheme can be applied. The electrical signal and an additive noise from the photodetector pass through the narrowband IF filter with central frequency ω_c and then through an ideal limiter accompanied by a zone filter, as illustrated in Figure 2.16. The signal at the limiter output has the constant amplitude S_ℓ, but its crossing points with the t-axis can differ from those in the sent DPSK signal, because of the noise influence.

The output random process from the limiter is led to the product modulator, where it is multiplied by its own value but delayed for one digit interval. Thus, the

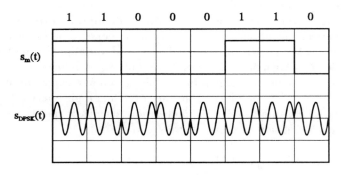

Figure 2.15 The waveforms of source digital signal and corresponding DPSK signal.

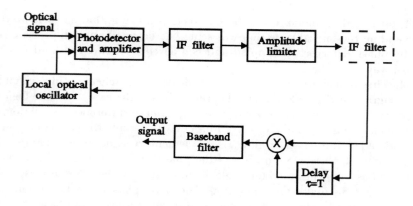

Figure 2.16 The general scheme of optical DPSK receiver.

elementary signal from a given digit interval is multiplied by the elementary signal from the preceding digit interval. The product is filtrated in the low-pass filter, and after that the sampling and decision are made.

The circuit for multiplication and the baseband filter from Figure 2.16 have the role of phase detector. After the multiplication of the incoming with the delayed signal and after the low-pass filtration, the following signal is obtained:

$$S^2 \cos[\omega_c(t - nT_d)] + \varphi_n]\cos[\omega_c(t - nT_d - \tau) + \varphi_{n-1}] = Z \cos(\varphi_n - \varphi_{n-1}) \qquad (2.123)$$

where it is assumed that the delay time, τ, is equal to the digit-interval duration, T_d, and that the digit interval is an integer product of the carrier wave period. The parameter Z presents a constant of phase detector.

Equation (2.123) shows that a true copy of a modulating signal is obtained at the output of the phase detector, but only if there is no noise. The additive noise influence can be considered by assuming that a false phase jump has been observed in a mth digit interval of transmitted signal. At the output of the phase detector, there will be the copy of the transmitted signal, which will contain not only the false jump in the mth digit interval, but also the false jump in the $(m + 1)$ digit interval. Hence, the error in one received elementary signal generates, in general, a couple of errors in subsequent digit intervals of detected DPSK signal [8].

2.7.2 Differentially Coherent Detection of PSK Signals

The DCPSK detection scheme will be analyzed using Figure 2.16 as an illustration. It can be assumed that the filtrated DPSK signal, $s(t)$, and filtrated additive noise, $n(t)$, are coming to the input of the phase detector. According to common assumption, we will consider that the noise has the zero mean value and that its variance σ^2 does not depend on the logical state that was sent. The detected random process, coming from the phase detector, is sampled and compared with the threshold $b = 0$ in a standard way.

The situation characteristic for a DCPSK detection scheme is illustrated by the phaser diagram in Figure 2.17. The diagram refers to the case when the logical "0" state $(a_n = 1)$ was sent. Thus, there is not the phase difference in relation to the preceding elementary signal $(\Delta\varphi = 0)$. Phaser S_0 represents the elementary signal in the considered digit interval, while S_{-1} is the elementary signal received in the preceding digit interval. Phaser S_0 consists of the phaser of useful signal S and the noise phaser N. Phaser S_{-1} is rotated for angle φ_{-1} in relation to the referent axis, while the phasers S_0 and S_{-1} make the angle $\Delta\varphi_n$. The noise phaser can be decomposed in components X and Y, which are in phase and in quadrature with phaser S_{-1}, respectively. The decision threshold is denoted by the dashed line. (In the absence of noise, the phasers S_0 and S_{-1} will cover the signal phaser S, since Figure 2.17 concerns the case when the "0" state was sent.) The phase shift, $\Delta\varphi_n$, which appeared due to noise influence, must be below $\mp\pi/2$, to prevent the situation when the decider "means" that the sent phase difference between phasers S_0 and S_{-1} was $\Delta\varphi = \pi$.

It can be seen from Figure 2.17 that a false decision will be made if the phaser X is greater than the component $S \cos \varphi$. Since the even component $x(t)$ of noise is the normal process with zero mean value and variance σ^2, the mentioned condition can be used to evaluate the conditional error probability that phaser S_{-1} differs from phaser S for the angle φ, or that $X > S \cos \varphi$. It can be expressed as

$$P(e/\varphi) = P(X > S \cos \varphi) \tag{2.124}$$

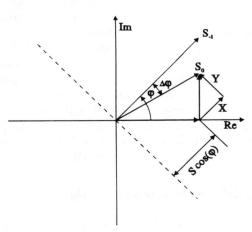

Figure 2.17 The phasers diagram corresponding to binary signals detection in DCPSK receiver.

where e associates to "error." It is more suitable to use the following normalized relation instead (2.124):

$$P(e/\varphi) = P\left(\frac{X}{\sqrt{2}\sigma} > \frac{S}{\sqrt{2}\sigma} \cos \varphi\right) \qquad (2.125)$$

In accordance with the coherent detection of PSK signals, this conditional error probability can be evaluated by

$$P(e/\varphi) = 0.5 \ \mathrm{erfc}\left(\frac{S \cos \varphi}{\sqrt{2}\sigma}\right) \qquad (2.126)$$

Equation (2.126) can take a more suitable form, by use of $\mathrm{erfc}(z) = 1 - \mathrm{erf}(z)$ and

$$\mathrm{erf}(z) = \frac{2}{\sqrt{\pi}} \int_0^z \exp(-z^2)dz \qquad (2.127)$$

The integral in (2.127) can be replaced by the expansion [4]

$$\int_0^z \exp(-z^2)dz = \sum_{k=0}^{\infty} \frac{(-1)^k}{k!(2k+1)} z^{2k+1} \qquad (2.128)$$

so we have that the conditional error probability is

$$P(e/\varphi) = \frac{1}{2} - \sum_0^{\infty} C_k\left(\frac{S \cos \varphi}{\sqrt{2}\sigma}\right)^{2k+1} \qquad (2.129)$$

where C_k is the coefficient

$$C_k = \frac{1}{\sqrt{\pi}} \frac{(-1)^k}{k!(2k+1)} \qquad (2.130)$$

The total error probability in a DPSK system can be found by averaging $P(e/\varphi)$ over all possible values of angle φ (in the region from $-\pi$ to $+\pi$), or

$$P_{eDCPSK,2} = \int_{-\pi}^{\pi} w(\varphi)P(e, \varphi)d\varphi \qquad (2.131)$$

The function $w(\varphi)$ corresponds to the Rice probability density function for instantaneous phase [Appendix D, (D. 16)]. The Rice function can be expressed in the form

$$w(\varphi) = \frac{1}{2\pi} \exp\left(-\frac{S^2}{2\sigma^2}\right) + \frac{S \cos \varphi}{\sqrt{2\pi}\sigma} \exp\left(-\frac{S^2 \sin^2 \varphi}{2\sigma^2}\right) \cdot [1 - P(e/\varphi)] \quad (2.132)$$

Thus, after substitution of (2.129) and (2.132) into (2.131), we have

$$P_{eDCPSK,2} = \int_{-\pi}^{\pi} \left\{ \frac{1}{2\pi} \exp\left(-\frac{S^2}{2\sigma^2}\right) + \frac{S \cos \varphi}{\sqrt{2\pi}\sigma} \exp\left(-\frac{S^2 \sin^2 \varphi}{2\sigma^2}\right) \cdot [1 - P(e/\varphi)] \right\} P(e/\varphi) d\varphi$$
$$(2.133)$$

The integral in (2.133) is complicated, but it can be solved analytically, and the following solution is obtained:

$$P_{eDCPSK,2} = \frac{1}{4\pi} \exp\left(-\frac{S^2}{2\sigma^2}\right) \int_{-\pi}^{\pi} d\varphi = 0.5 \exp\left(\frac{S_h^2}{2\sigma_h^2}\right) \quad (2.134)$$

The subscript h was introduced again to unify the denotation in the expressions for error probabilities of different heterodyne detection schemes. Equation (2.134) is rather simple and shows that there is a small difference between coherent (CPSK) and differentially coherent (DCPSK) detection of binary PSK signals for higher SNRs.

It is quite understandable that the error probability for a DCPSK detection scheme is higher than the error probability for a CPSK detection scheme, since in DCPSK detection there is noise influence not only on the incoming information signal but also on the local optical oscillator signal. Hence, we can expect the sensitivity of a DCPSK optical receiver to be 3 dB below the sensitivity value of a CPSK receiver. But the calculations show that the receiver sensitivity degradation is not 3 dB but slightly above 1 dB, which is very favorable for DCPSK receivers [5, 7].

As for an M-value DCPSK system, in order to find the receiver sensitivity, it can be assumed that elementary signals with the equal initial phases $\varphi_{n-1} = \varphi_n$ were sent in the digit intervals $(n - 1)$ and n, respectively. In such a way, there are two random processes in the nth digit interval, which act at the ideal phase detector input. These processes are given by the equations

$$s_n(t) + n_n(t) = [S + x_n(t)]\cos(\omega_c t - \varphi_n) - y_n(t)\sin(\omega_c t - \varphi_n) \quad (2.135)$$

$$s_{n-1}(t) + n_{n-1}(t) = [S + x_{n-1}(t)]\cos(\omega_c t - \varphi_{n-1}) - y_{n-1}(t)\sin(\omega_c t - \varphi_{n-1}) \quad (2.136)$$

with the assumption that the noises in subsequent digit intervals are statistically independent. Thus, according to the phaser diagram in Figure 2.17, the equation for the phase difference of phasers S_n and S_{n-1} can be written [1]

$$\Delta\varphi = \frac{y_n(t)}{S + x_n(t)} - \frac{y_{n-1}(t)}{S + x_{n-1}(t)} \tag{2.137}$$

A false decision will be made if the random phase shift, according to (2.137), is above the value $\mp \pi/M$, where M is the number of the possible states of the phase difference. The total error probability is given as

$$P_{eDCPSK,M} = 1 - \int_{-\pi/M}^{\pi/M} w(\Delta\varphi)d(\Delta\varphi) \tag{2.138}$$

where $w(\Delta\varphi)$ is the probability density function of the random phase difference $\Delta\varphi$. The random variable $\Delta\varphi$ presents the difference of two random processes, which submits to the Rice distribution (D.16), where $w_n(\varphi_n) = w_{n-1}(\varphi_{n-1}) = w(\varphi)$. It is well known [6] that the following equation can be written for probability density functions:

$$w(\Delta\varphi) = \int_{-\pi}^{\pi} w(\varphi)w(\varphi + \Delta\varphi)d\varphi \tag{2.139}$$

The integral in (2.138) should be solved by using (D.16) and (2.139), but the solution cannot be found in a closed form. Thus, the solution must be found numerically. Equation (2.138) can be approximated by a simpler one to make the comparison with the corresponding expressions for other modulation/detection schemes. Such an approximation, valid for high SNRs, for error probability in an M-valued DCPSK system is [14]

$$P_{eDCPSK,M} \simeq \exp\left[-\frac{S^2 \sin^2(\pi/2M)}{2\sigma^2}\right] \tag{2.140}$$

By comparing (2.14) with (2.112), which refers to the error probability in an M-valued PSK system, it can be seen that the exponent in (2.140) should be increased by the factor

$$q = \frac{\sin^2(\pi/M)}{2 \sin^2(\pi/2M)} \tag{2.141}$$

to reach the error probability in a CPSK system. It can also be concluded that a 4-DCPSK optical receiver requires a 2.3-dB-higher optical signal power than a 4-PSK optical receiver for the same error probabilities; for an 8-DCPSK optical receiver, that increase is 2.8 dB. For the values of M above 8, the increase in required optical signal power tends asymptotically to value 3 dB. Hence, 3 dB is the measure of the sensitivity degradation of an ideal DCPSK optical receiver.

2.7.3 Influence of Phase Noise on Differential Detection of PSK Signals

Keeping in mind (2.51), (2.52), and (2.56), and the general scheme of a DCPSK optical receiver, we can express the phase error that appears in the receiver as the difference between the phase shifts in the considered digit interval and the preceding one, or

$$\Delta \vartheta_n(t) = \vartheta_n(t) - \vartheta_n(t - T_d) \tag{2.142}$$

This difference of phase shifts is, in fact, an additional component of the phase difference between the phasers S_n and S_{n-1}, which is given by (2.137). Actually, the total phase difference can be written as

$$\Delta \varphi_t = \Delta \varphi + \Delta \vartheta_n \tag{2.143}$$

where $\Delta \varphi$ is the phase difference in the absence of phase noise, according to (2.137), while $\Delta \vartheta_n$ is the phase shift due to phase noises, given by (2.142). The total error probability can be found by averaging all possible values of $\Delta \vartheta_n$, knowing the probability density function of $\Delta \vartheta_n$, or

$$P_{E, DCPSK/M} = \int_{-\pi/M}^{\pi/M} P_{eDCPSK, M}(\Delta \vartheta_n) w(\Delta \vartheta_n) d(\Delta \vartheta_n) \tag{2.144}$$

where $P_{e, DCPSK, M}(\Delta \vartheta_n)$ represents the error probability evaluated by (2.138) for the argument $\Delta \varphi_t$ instead of $\Delta \varphi$. Since the random variable $\Delta \vartheta_n$ is the difference between two Gaussian processes with zero mean values and the variance σ_n^2, the probability density functions for $\Delta \vartheta_n$ can be found as

$$w(\Delta \vartheta_n) = \int_{-\pi}^{\pi} w(\vartheta_n) \, w(\vartheta_n + \Delta \vartheta_n) d\vartheta_n = \frac{1}{2\pi\sigma_n^2} \int_{-\pi}^{\pi} \exp\left(-\frac{2\vartheta_n^2 + 2\vartheta_n \Delta \vartheta_n + \Delta \vartheta_n^2}{2\sigma_n^2}\right) d\vartheta_n \tag{2.145}$$

The integral in (2.145) can be solved analytically, so the probability density function for $\Delta \vartheta_n$ takes the form

$$w(\Delta \vartheta_n) = \frac{1}{2\sqrt{\pi} \, \sigma_n} \exp\left(-\frac{\Delta \vartheta_n^2}{4\sigma_n^2}\right) \text{erf}\left(\frac{3\pi + \Delta \vartheta_n}{2\sigma_n}\right) \tag{2.146}$$

The error probability according to (2.144) can be found numerically only for the desired values of noise variance σ_n^2 and defined number M. It can be shown that the calculated values for 2-DPSK detection scheme are between the values corresponding to FSK and PSK detection schemes, that is, the error probability in a real DPSK system is below the error probability for a real FSK system but above the error probability for a real PSK lightwave system [7, 14, 15, 16].

2.8 HOMODYNE DETECTION OF OPTICAL SIGNALS

Homodyne detection presents a special case of heterodyne detection, where the frequency of local optical oscillator is equal to the frequency of an incoming information optical signal. Thus, (2.8) can be applied for homodyne detection if we set that $\Delta\omega = 0$. The homodyne detection scheme can be applied both for detection of ASK and PSK signals. In such detection schemes, there must be not only frequency synchronization but signal phase synchronization as well. The detection of FSK signals can be performed by homodyne methods only in the special optical receiver with two phase-controlled local optical oscillators. That realization is very complex and not actual in the near future.

The homodyne optical receiver contains a baseband amplifier and low-pass filter following the optical mixer and photodetector, instead of an IF amplifier and IF filter, as in the heterodyne optical receiver scheme (see Figure 2.3 and 2.11). Such structure of the optical receiver brings two important advantages, compared with a heterodyne detection scheme. First, the frequency bandwidth in a homodyne optical receiver is only half the heterodyne receiver bandwidth; hence, the total power of noise is half the noise power in a heterodyne receiver. At the same time, the homodyne receiver sensitivity is increased 3 dB compared with heterodyne receiver sensitivity. Second, the boundary frequency of the photodetector response in a homodyne detection scheme is nearly equal to the frequency bandwidth of the modulating signal, while heterodyne detection requires a photodetector boundary frequency several times larger than the frequency bandwidth of modulating signal [17–20].

To find the error probability in homodyne detection schemes of ASK and PSK signals, we will use (2.7) for the optical power at the photodetector. For $\omega_s = \omega_L$ and for $P_s \ll P_L$ and $\Delta\varphi(t) \simeq 0$, the value of the current signal becomes

$$i(t) = \Re M\left[P_L + 2\sqrt{P_s(t)P_L} \right] \qquad (2.147)$$

where $\Re$ is the photodiode responsivity, and M is the avalanche gain coefficient. Hence, the value of the current signal is nearly equal to the current signal for heterodyne detection. Since the bandwidth of the baseband filter of a homodyne receiver is equal to half the IF filter bandwidth in a heterodyne receiver, we have

$$\sigma_H^2 = \sigma_h^2/2 \qquad (2.148)$$

where subscript $_H$ refers to the homodyne detection scheme and subscript $_h$ to the heterodyne detection.

The value of the sampled signal before the decision process in an ASK receiver and the value of the optimum threshold for that case are given by (2.40) and (2.42), respectively; they can be applied to homodyne detection, as well. Hence, (2.43) can be used to evaluate error probability in homodyne detection schemes of ASK signals, and the following relation can be obtained:

$$P_{HASK,2} = 0.5 \text{ erfc}\left(\frac{S_H}{2\sqrt{2}\sigma_H}\right) \tag{2.149}$$

By using the same method, (2.99) for error probability in a heterodyne detection scheme of PSK signals can be modified, to obtain the expression for error probability in a PSK homodyne detection scheme. Thus, we have

$$P_{HPSK,2} = 0.5 \text{ erfc}\left(\frac{S_H}{\sqrt{2}\sigma_H}\right) \tag{2.150}$$

The advantage of homodyne detection in terms of the value of boundary frequency of photodetector response plays a great role for very high transmission rates (over 1 Gb/s). For these transmission rates, speed PIN photodiodes or APDs with standard design can be used in homodyne receivers, while very speed photodiodes with nonstandard design must be used in heterodyne optical receivers [18, 19, 21].

However, in the practical realization of homodyne receivers, great difficulties related to the preservation of frequency and phase stability of local optical oscillators are being encountered. The phase of the local optical oscillator must follow the phase of the incoming optical signal. If the phases of these two signals are not synchronized, the output current signal becomes very close to zero value at every moment when the phases of the incoming optical signal and the local optical oscillator signal differ for $\pi/2$. But even if the phases are mutually synchronized, there will be the influence of the phase noise due to the quasi-coherent nature of light sources, which can lead to the serious degradation of the homodyne optical receiver characteristics.

The phase tracking is made by an optical PLL where the local optical oscillator follows the phase of the incoming optical signal and provides that the detected current signal will be in base frequency band, as in IM/DD lightwave systems. Another possibility for improved phase tracking in homodyne optical receivers scheme is the employment of the so-called "phase diversity technique," which will be discussed in Chapter 4.

It should be pointed out that the results related to the influence of phase noise on the error probability in ASK and PSK detection schemes and given by (2.58) and (2.119), respectively, can be applied in homodyne detection schemes as well. In that case, $\Delta\vartheta_n$ presents the phase error that appears in the process of phase tracking the incoming optical signal by the phase-controlled local optical oscillator.

Closing this chapter, we can summarize and compare receiver sensitivities of various methods of coherent and noncoherent detection, which are interesting for practical employment. This is given in the Table 2.1. The sensitivity of the receiver belonging to the IM/DD system plays a reference role.

To compare sensitivities of coherent and noncoherent detection schemes the approximation (2.35) has been applied. The second column in Table 2.1 presents the relative sensitivity improvement in comparison with the detection scheme from the previous row. The relative values of the receiver sensitivity in comparison with the reference system (IM/DD) sensitivity are displayed in the third column. The last column in Table 2.1 shows the optical receiver bandwidth through the photodiode boundary frequency that corresponds to the detection scheme.

As we can see, there is the range of about 30 dBm where the sensitivities are placed. At the same time, receiver bandwidth varies from the value B corresponding to the bandwidth of the IM/DD receiver to the value 2B corresponding to the heterodyne detection scheme. Photodiode boundary frequency f_b varies from the value approximately equal to the signal bandwidth B to several times higher value. The last column in Table 2.1 also contains remarks about receiver complexity for homodyne detection schemes. This table can be used as a quick reference for the comparison of various detection schemes in a coherent lightwave system.

Table 2.1
Comparative Review of Various Optical Detection Schemes

Modulation Method	Sensitivity Improvement	Sensitivity Relative to Reference Value	Receiver Bandwidth/ Photodiode Boundary Frequency
IM/DD	Reference sensitivity	0 dB	B/f_b
ASK heterodyne noncoherent	>12 dB improvement	<−12 dB	2B/more than $3f_b$
ASK heterodyne coherent	2.2 dB improvement	<−14.2 dB	2B/more than $3f_b$
FSK heterodyne noncoherent	0.8 dB improvement	<−15 dB	2B/more than $3f_b$
FSK heterodyne coherent	2.2 dB improvement	<−17.2 dB	2B/more than $3f_b$
2-value DCPSK heterodyne	1.8 dB improvement	<−19 dB	2B/more than $3f_b$
2-value PSK heterodyne	1.2 dB improvement	<−20.2 dB	2B/more than $3f_b$
4-value DCPSK heterodyne	0.7 improvement	<−20.9 dB	2B/more than $3f_b$
8-value DCPSK heterodyne	1.5 dB improvement	<−22.4 dB	2B/more than $3f_b$
4-value PSK heterodyne	0.8 dB improvement	<−23.2 dB	2B/more than $3f_b$
2-value PSK homodyne	No improvement	<−23.2 dB	B/f_b; more complex receiver
8-value PSK heterodine	2 dB improvement	<−25.2 dB	2B/more than $3f_b$
4-value PSK homodyne	1 dB improvement	<−26.2 dB	B/f_b; more complex receiver
8-value PSK homodyne	2 dB improvement	<−28.2 dB	B/f_b; more complex receiver

REFERENCES

[1] Lee, Y. W., *Statistical Theory of Communications*, New York: John Wiley and Sons, 1960.

[2] Naito, T., et al., "Optimum System Parameters for Multigigabit CPFSK Optical Heterodyne Detection System, *IEEE/OSA J. Lightwave Techn.*, LT-12 (1994), pp. 1835–1841.

[3] Personick, S. D., *Optical Fiber Transmission Systems*, New York: Plenum, 1981.

[4] Korn, G., and T. Korn, *Mathematical Handbook for Scientists and Engineers*, London: McGraw Hill, 1960.

[5] Kikuchi, K., et al., "Degradation of Bit Error Rate in Coherent Optical Communications Due to Spectral Spread of the Transmitter and the Local Oscillator," *IEEE J. Lightwave Tech.*, LT-2(1984), pp. 1024–1033.

[6] Sing, N., et al., "Performance of Heterodyne Coherent Optical Fiber Communication Systems in Presence of Shot Noise, LO Excess Noise, Laser Phase Noise and Time Jitter," *J. Optical Commun.*, 10(1989), pp. 48–53.

[7] Okoshi, T., et al., "Computation of Bit-Error Rate of Various Heterodyne and Coherent Type Optical Communications Schemes," *J. Optical Commun.*, 2(1981), pp. 89–96.

[8] Witerbi, A. J., *Principles of Coherent Communications*, New York: McGraw Hill, 1966.

[9] Lachs, G., et al., "Sensitivity Enhancement Using Coherent Heterodyne Detection," *IEEE/OSA J. Lightwave Techn.*, LT-12(1994), pp. 1036–1041.

[10] Garreth, I., and G. Jacobsen, "Theoretical Analysis of Heterodyne Optical Receivers for Transmission Systems Using (Semiconductor) Laser with Nonnegligible Linewidth," *IEEE J. of Lightwave Techn.*, LT-4(1986), pp. 323–334.

[11] Schwartz, M., et al., *Communication Systems and Techniques*, New York: McGraw-Hill, 1966.

[12] Huang, S., and J. Yan, "Power Penalty Due to Intersymbol Interference in Optical Phase Diversity FSK Systems," *J. Optical Commun.*, 2(1991), pp. 59–60.

[13] Chikama, T., et al., "Modulation and Demodulation Techniques in Optical Heterodyne PSK Transmission Systems, *IEEE/OSA J. Lightwave Techn.*, LT-9(1991), pp. 309–322.

[14] Cahn, C. R., "Performance of Digitial Phase-Modulation Communications Systems," *Ramo-Wooldrige Rept.*, M1/0-905, April 1959.

[15] Bischel, H., and F. Derr, "Chromatic and Polarization Dispersion Limitations in Coherent Optical BPSK and QPSK Systems," *J. Optical Commun.*, 12(1991), pp. 42–46.

[16] Tsao, H. W., et al., "The Performance Analysis of Decision-Driven Optical Phase Locked Loop With Loop Delay," *J. Optical Commun.*, 11(1990), pp. 70–75.

[17] Glance, B., "Performance of Homodyne Detection of Binary PSK Optical Signals," *IEEE/OSA J. of Lightwave Techn.*, LT-4(1986), pp. 228–235.

[18] Kazovsky, L. G., "Optical Heterodyning Versus Optical Homodyning: A Comparison," *J. Optical Commun.*, 6(1985), pp. 18–24.

[19] Fleischmann, M., and J. Frantz, "Optimization of Coherent Optical Homodyne Systems," *J. Optical Commun.*, 9(1988), pp. 72–76.

[20] Dery, R. J., et al., "High-Spread Heterodyne Operation of Monolithically Integrated Balanced, Polarization Diversity Photodetectors," *Electron. Letters*, 28(1992), pp. 2332–2334.

[21] Davidson, F. M., and C. T. Field, "Coherent Homodyne Optical Communications Receivers With Photorefractive Optical Beam Combiners," *IEEE/OSA J. Lightwave Techn.*, LT-12(1994), pp. 1207–1223.

Chapter 3

Optical Transmitters for Coherent Lightwave Systems

3.1 INTRODUCTION

The optical transmitter in a coherent lightwave system performs the functions of generating a coherent optical wave and impressing information on the wave by modulation. The use of a laser as the light source is essential to obtain the necessary degree of optical coherence. There are several ways of exciting laser media to obtain stimulated emission, or *laser action*. Of these, by far the most widely used techniques are optical and electrical excitation. The output light wave can therefore be modulated by controlling the excitation of the laser (for example, by modulating the electrical drive current) or by using a separate light modulator external to the laser itself. Semiconductor lasers are generally favored for coherent optical lightwave systems because of their small geometry and the relatively simple and efficient internal modulation obtained by varying the electrical drive current.

In this chapter we will consider the main characteristics of a coherent optical transmitter, in terms of the source excitation and optical radiation modulation. We will, of course, consider the methods employed to regulate the laser operating regime and to stabilize the output optical radiation parameters (carrier wavelength, spectral width, initial phase). The very beginning of this chapter will describe the general modulation methods; after that, we will consider the main characteristics of semiconductor lasers and explain the nature and characteristics of laser noise.

3.2 MODULATION OF OPTICAL RADIATION

The modulation of optical radiation refers to the variations of essential parameters of the carrier optical wave (amplitude, frequency, phase) in accordance with variations of the information signal parameters. The information signal is a modulating signal that is carried over an optical fiber line and then received and recognized at the optical receiver. Because the carrying optical frequencies, which belong to the visible or near infrared regions, are in the region of 10^{14} to 10^{15} Hz, it is basically possible to modulate the carrying optical wave with a signal having a bandwidth of even 10^{12} Hz.

The applied modulation method depends on the concrete requirements and the system's character. Optical modulators, in general, should have the following features: wide modulation bandwidth, linear modulation characteristics, large dynamic range, and low energy consumption.

There are two general methods for optical radiation modulation: internal modulation and external modulation. Internal modulation includes the influence on the optical signal still in the process of optical-wave generation; thus, it is simultaneous with the light-source excitation. External modulation means the influence on a generated referent wave after it leaves the laser resonator. That influence is made by an information modulating signal in a device called an *optical modulator*. External modulation is based on the physical effects, such as electro-optic effect, acousto-optic effect, magneto-optic effect, and so on.

The types of modulation are amplitude modulation, frequency modulation, phase modulation, polarization modulation, and intensity modulation. Intensity modulation also can be assumed to be a special case of amplitude modulation. Intensity modulation is used in IM/DD lightwave systems and will not be considered in this chapter. It is mentioned only as an illustration of the considered effects. Polarization modulation is a modulation method that will become attractive if the problems related to the polarization state preservation during signal propagation are solved.

The modulation of an optical signal can be made directly by information signal or by a signal with subcarrier frequency. In the latter case, the information signal modulates the electrical signal from the medium frequency band (corresponding to the IF band in the optical receiver), then that modulated signal modulates the carrier optical frequency.

The analog modulation methods are characterized by continuous variations of the relevant optical wave parameter (amplitude, frequency, phase, polarization state) in accordance with the continuous variations of the information signal amplitude. Thus, the continuous range of the output optical wave parameter corresponds to the continuous range of the input modulating signal. The digital modulation methods (ASK, FSK, PSK, and digital polarization modulation) are characterized by the rule that a discrete state of the relevant optical wave parameter corresponds to a discrete value of the modulating signal amplitude. These methods will be the main subject of the next part of this chapter.

The different physical principles are at the base of these modulation methods. ASK modulation can be realized by using electro-optical, piezoelectric, or acousto-optical effects. Besides, the dependence of the light absorption on the parameter of the optical wave and the kind of material can be used for ASK modulation, especially in semiconductor lasers. But the most simple ASK modulation method is by laser pump, where the amplitude of the referent optical wave is changed under deviations of the pump current.

The Zeeman and Stark effects are efficient for realization of the FSK modulation method, but FSK optical signals can be obtained by employing other effects, such as electro-optical, magneto-optical and acousto-optical effect, as well. It should be noted that the modulation elements for realization of ASK and FSK modulations can also be placed in the laser resonator, where they can influence the optical resonator length and the resonator loss. In such a case, the internal modulation is, in fact, transformed into modulation by pumping. As for the applied effects for digital phase modulation, we can say that these methods are like the FSK methods.

The next subsection will briefly explain the basic principles of the modulation of referent optical waves. Such explanation is necessary for a better understanding of the operation of optical modulators.

3.2.1 Refractive Index Features in Crystals

The large number of modulation methods are based on the refractive index characteristics or, rather, on the fact that the refractive index is dependent on the geometrical parameters of crystals and the parameters of an external field. Thus, some fundamental theoretical facts about electro-optics are essential for the design of optical modulators.

Under the action of an electric field on an anisotropic medium, the density of the electrical energy in the medium is expressed as [1]

$$w = \frac{\mathbf{E}\mathbf{D}}{2} \tag{3.1}$$

where $\mathbf{E}$ is the electric field vector, and $\mathbf{D}$ is the electric induction vector. In an anisotropic crystal the following tensor relation is valid for these vectors:

$$\mathbf{D} = \overset{\Rightarrow}{\epsilon}\mathbf{E} \tag{3.2}$$

The tensor of dielectric permittivity, $\overset{\Rightarrow}{\epsilon}$, can be expressed by its own components as

$$\overset{\Rightarrow}{\epsilon} = \begin{pmatrix} \epsilon_{xx} & \epsilon_{xy} & \epsilon_{xz} \\ \epsilon_{yx} & \epsilon_{yy} & \epsilon_{yz} \\ \epsilon_{zx} & \epsilon_{zy} & \epsilon_{zz} \end{pmatrix} \tag{3.3}$$

Equation (3.2) can be rewritten as

$$D_k = \sum_m \epsilon_{km} E_m \tag{3.4}$$

where the subscripts k and m denote the coordinates $x, y,$ or $z,$ related to the referent system. Hence, the density of the electrical energy in the crystal takes the form

$$w = \frac{1}{2} \sum_k \sum_m \epsilon_{km} E_k E_m \qquad k, m = x, y, z \tag{3.5}$$

In general, the tensor of dielectric permittivity has six independent components, because $\epsilon_{km} = \epsilon_{mk}$ [1], so the density of the electrical energy can be written as

$$2w = \epsilon_{xx} E_x^2 + \epsilon_{yy} E_y^2 + \epsilon_{zz} E_z^2 + 2\epsilon_{yz} E_y E_z + 2\epsilon_{xz} E_x E_z + 2\epsilon_{xy} E_x E_y \tag{3.6}$$

Equation (3.6) can be simplified by corresponding coordinate transformation. Thus, the last three terms vanish, and only the terms related to the main axes stay. Those three new axes are called the *principal dielectric axes*. Equation (3.6) takes the following form in a new coordinate system

$$2w = \epsilon_x E_x^2 + \epsilon_y E_y^2 + \epsilon_z E_z^2 \tag{3.7}$$

In this new coordinate system, the tensor of the dielectric permittivity has a diagonal form, so the relation between the electric field and the electric induction components becomes

$$\begin{pmatrix} D_x \\ D_y \\ D_z \end{pmatrix} = \begin{pmatrix} \epsilon_x & 0 & 0 \\ 0 & \epsilon_y & 0 \\ 0 & 0 & \epsilon_z \end{pmatrix} \cdot \begin{pmatrix} E_x \\ E_y \\ E_z \end{pmatrix} \tag{3.8}$$

The density of the electric field energy in the crystal takes the form

$$2w = \frac{D_x^2}{\epsilon_x} + \frac{D_y^2}{\epsilon_y} + \frac{D_z^2}{\epsilon_z} \tag{3.9}$$

The above relation defines the ellipsoidal surface in the $[D_x, D_y, D_z]$ space. By using the substitutions

$$\frac{D_x}{\sqrt{2w}} = x, \quad \frac{D_y}{\sqrt{2w}} = y, \text{ and } \frac{D_z}{\sqrt{2w}} = z \tag{3.10}$$

and the corresponding relations between the refractive index and dielectric permittivity,

$$n_x = \sqrt{\epsilon_x}, \; n_y = \sqrt{\epsilon_y}, \text{ and } n_z = \sqrt{\epsilon_z} \qquad (3.11)$$

(3.9) becomes

$$\frac{x^2}{n_x^2} + \frac{y^2}{n_y^2} + \frac{z^2}{n_z^2} = 1 \qquad (3.12)$$

Equation (3.12) is the canonic equation for an ellipsoid, with main axes covering the coordinates x, y, and z. This ellipsoid is called the *ellipsoid of wave normals* [1], the *principal ellipsoid* (which we will use), or the *optical indicatrix*. The shape of the ellipsoid depends on its semi-axes, defined by the refractive indexes n_x, n_y, and n_z. Thus, there are several cases:

- If the refractive indexes are equal in all coordinate directions $(n_x = n_y = n_z = n)$, the ellipsoid takes the form of a ball, and the crystal is called *isotropic dielectric*.
- When $n_x = n_y = n_1$ and $n_z = n_2$ $(n_1 \neq n_2)$, the ellipsoid is obtained by ellipse rotation around the z-axis, and the crystal is called *one-axis dielectric*.
- If $n_x \neq n_y \neq n_z$, the ellipsoid is called *two-axes dielectric*.

The light ray propagating in an optical one-axis medium is generally divided into *ordinary ray* and *extraordinary ray*, with except the ray propagating parallel to the optical axis. The refractive coefficient of the ordinary ray is denoted by $n_x = n_y = n_0$, while the refractive coefficient of the extraordinary ray is denoted by $n_z = n_e$. This different denotation is because both the refractive indexes and the phase velocities of the ordinary and extraordinary rays are different. The ordinary ray is polarized perpendicularly to the plane cutting the crystal axis, while the extraordinary ray is polarized parallel to that plane.

The observed effect when two rays are propagating with different phase velocities in one direction is called the *birefringence effect*. This effect is the base for the operation of electro-optical and magneto-optical modulators. The birefringence effect can be used in such modulators because the refractive indexes of the principal ellipsoid depend on the external electric or magnetic fields. The principal ellipsoid is deformed under the influence of the external field and changes its position in relation to the main axes or in relation to the direction of ray propagation.

A similar effect can also appear under the influence of the acoustic waves caused by mechanical force. In such a case, we have a *photoelasticity effect*, which can be used for optical radiation modulation. But the modulation rate is rather small, so the application of the photoelasticity effect in coherent communications systems is limited.

3.2.1.1 Linear Electro-Optical Effect

Equation (3.12) of the principal ellipsoid can be written in the form

$$a_{10}x^2 + a_{20}y^2 + a_{30}z^2 = 1 \tag{3.13}$$

where a_{j0} ($j = 1, 2, 3$) are the reciprocal values of the refractive index squares in corresponding axes. Subscript 0 means that the coefficients are related to the case without the external electric field. Under the action of the external electrical field, the ellipsoid will be deformed. The main axes of a newly formed ellipsoid will not cover the axes of the initial ellipsoid, so the equation of the principal ellipsoid in the former coordinates can be written as

$$a_1x^2 + a_2y^2 + a_3z^2 + 2a_4yz + 2a_5xz + 2a_6xy = 1 \tag{3.14}$$

The linear electro-optical effect, or *Pockel's effect*, is characterized by the linear dependence of the ellipsoid coefficients on the external electric field. Thus, it is valid that

$$a_j - a_{j0} = r_{j1}E_x + r_{j2}E_y + r_{j3}E_z, j = 1, 2, \ldots, 6 \tag{3.15}$$

In (3.15) it is assumed that the coefficients a_{j0} for $j > 3$ are equal to zero. The coefficients r_{jn} ($j = 1, \ldots, 6; n = 1, 2, 3$) form a 3-by-6 dimension matrix of electro-optical coefficients. Some of the matrix elements can be equal to zero, while some of them possess the same values.

So far, a few types of the crystals have been examined for application in electro-optical modulators. It has been shown that the most suitable characteristics belong to the crystals in the groups $\overline{4}2m$, $\overline{4}3m$, and 3m (GaAs belongs to group $\overline{4}3m$, while LiNoO$_3$ belongs to group 3m). The group $\overline{4}2m$ is not quite suitable for realization of the opto-electronic integrated versions, so application of the other two groups is preferable.

The principles of electro-optical modulation can be easily explained by the example of a crystal from the $\overline{4}2m$ group, because the matrix of the electro-optical coefficients belonging to this crystal is quite simple. In fact, only coefficients $r_{52} = r_{41}$ and r_{63} are different from zero. Besides, the coefficients a_{10} and a_{20} are equal; thus, the equation of the principal ellipsoid takes the form

$$a_{10}(x^2 + y^2) + a_{30}z^2 + 2r_{41}(E_xyz + E_yzx) + 2r_{63}E_zxy = 1 \tag{3.16}$$

If the electric field acts along the z-axis, corresponding to the optical axis of the crystal, or if $\bar{E}_x = E_y = 0$ and $E_z = E$, (3.16) becomes

$$a_{10}(x^2 + y^2) + a_{30}z^2 + 2r_{63}Exy = 1 \tag{3.17}$$

The canonical form of the ellipsoid is obtained by introducing the new coordinates x' and y', which are rotated around the z-axis for $\pi/4$, while the z-axis stays unchanged. Then (3.17) takes the form

$$(a_{10} - r_{63}E)x'^2 + (a_{10} + r_{63}E)y'^2 + a_{30}z^2 = 1 \tag{3.18}$$

It can be concluded from (3.17) and (3.18) that the cross-section of the principal ellipsoid with plane $z = 0$ is a circle if the electric field is absent. In the presence of the electric field, the cross-sectional area is an ellipse with the semi-axes

$$n_{x'} = \frac{1}{\sqrt{a_{10} - r_{63}E}} \simeq n_0 + \frac{1}{2}n_0^3 r_{63}E \tag{3.19}$$

$$n_{y'} = \frac{1}{\sqrt{a_{10} + r_{63}E}} \simeq n_0 - \frac{1}{2}n_0^3 r_{63}E \tag{3.20}$$

Equations (3.19) and (3.20) show how the refractive index values along the axes x' and y' depend on the external electric field.

The incident wave is, due to the birefringence effect, split into two components. One component is polarized along the x'-axis and has the phase velocity $v_{x'} = c/n_{x'}$; the other is polarized along the y'-axis and has the phase velocity $v_{y'} = c/n_{y'}$. It can be shown that, in such a way, phase modulation of the light polarized along the x'-axis (or the y'-axis) can be obtained. The delay of the outgoing wave depends on the external electric field and can be evaluated as

$$\delta_{x'} = \frac{2\pi}{\lambda}(n_{x'} - n_0)\ell = \pi n_0^3 r_{63} E\ell/\lambda \tag{3.21}$$

where λ is the wavelength of the light signal, and ℓ is the length of the ray path along the optical axis of the crystal. The phase delay of the wave polarized along the y'-axis can be determined in a similar way. This phase delay has the same value but the opposite sign as the delay expressed by (3.21).

Since the electric field acts along the z-axis, then the product $E\ell$ is voltage acting on the crystal. The phase shift between the waves polarized along the x' and the y'-axis now can be expressed as

$$\Delta\Phi = \frac{2\pi}{\lambda}(n_{x'} - n_{y'})\ell = 2\pi n_0^3 r_{63} U/\lambda \tag{3.22}$$

Since the optical signal will have elliptical polarization, the ellipse equation, for a sine voltage, for example, in the main coordinates x and y has the form

$$\frac{E_{sx}^2}{E_s \cos^2(\Delta\Phi/2)} + \frac{E_{sy}^2}{E_s \sin^2(\Delta\Phi/2)} = 1 \tag{3.23}$$

where E_s is the intensity of the electric field belonging to the optical signal.

The phase difference given by (3.22) becomes zero in the absence of direct voltage. At the same time, the polarization of the light at the output of the crystal is the same as the polarization of the light at the crystal input. The phase difference increases with the increase in the direct voltage. When $\Delta\Phi = \pi/2$, the output light will have circular polarization; for $\Delta\Phi = \pi$, the output polarization will be perpendicular to the input polarization. In such a way, the polarization of the input light radiation is changed proportionally to the direct voltage. The discrete variations of the direct voltage lead to the discrete variations of the polarization states in the output optical signal, which is the base of polarization modulation.

To convert the polarized modulated signal to an amplitude-modulated signal, the output signal must be passed through an analyzer, as illustrated in Figure 3.1. The relation between the direct voltage and the amplitude of the output optical signal can be obtained by following the preceding considerations, but it is not necessary to do so here because our subjects of interest are the digital modulation methods. Thus, we can say that the discrete values of polarization, defined by the angle between the vector E_s and the x-axis in Figure 3.1, correspond to the discrete values of the y-component of this vector, propagating through the analyzer. Accordingly, the

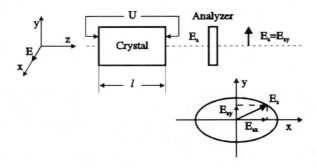

Figure 3.1 The application of the electro-optical effect for the modulation of light signal.

y-component of the signal electric field vector presents the digital modulated light signal.

3.2.1.2 Quadratic Electro-Optical Effect

The quadratic electro-optical effect, or *Kerr's effect*, is characterized by square-law dependence of the refractive index in a crystal on the intensity of an external electric field. This effect is the largest for the crystals from group m3m, so it is the most suitable to consider in such a situation.

Under the influence of the electric field directed along the z-axis, the crystals from the m3m group become one-axial, with the optical axis directed along the z-axis. The canonical form of the principal ellipsoid equation in an x,y,z coordinate system has the form

$$\left(\frac{1}{n_0^2} + h_{12}E^2\right)x^2 + \left(\frac{1}{n_0^2} + h_{12}E^2\right)y^2 + \left(\frac{1}{n_0^2} + h_{11}E^2\right)z^2 = 1 \qquad (3.24)$$

where h_{11} and h_{12} are the electro-optical coefficients characterizing the quadratic electro-optical effect.

From (3.24) it follows that the crystal is transformed from the isotropic form to the one-axial form, with the refractive indexes

$$n_x = n_y = n_0(1 - n_0^2 h_{12}E^2/2) \qquad (3.25)$$

$$n_z = n_0(1 - n_0^2 h_{11}E^2/2) \qquad (3.26)$$

If the light is polarized in the x (or y) direction and propagates along the z-axis, then the phase shift between the incoming and outgoing waves is

$$\delta = \pi\ell(n_0^3 h_{12}E^2)/\lambda \qquad (3.27)$$

When the light wave is polarized under the angle $\pi/4$ in relation to the z-axis and travels in the y direction, then there will be two different waves propagating in the crystal. These waves have equal amplitudes, but they are polarized along the axes x and y. If the crystal width is ℓ, the phase shift between these two waves is

$$\Delta\Phi = \frac{2\pi}{\lambda}(n_x - n_z)\ell = \frac{\pi\ell}{\lambda}(h_{11} - h_{12})n_0^3 E^2 \qquad (3.28)$$

The resultant vector of the electric field intensity at the crystal output will draw an ellipse. The ellipse deforms into a circle for phase difference $\Delta\Phi = \pi/2$, while for

$\Delta\Phi = \pi$ the ellipse becomes a line. The value of the phase shift is determined by the external voltage $U = E\ell$.

3.2.2 Amplitude Modulation of Optical Signals

The operating principle of an amplitude modulator can be explained by using Figure 3.1. Efficient amplitude modulation can be imposed if the crystal is between two polarization devices, whose principal planes are rotated to form some angle with the principal crystal axes. In the absence of a modulating signal (i.e., applied voltage) the light signal will not pass through the modulator. There will be birefringent refraction of the light signal, if the modulating signal is not zero. Thus, the phase difference between the ordinary and the extraordinary rays will appear.

The plane wave will be transformed into an elliptically polarized wave, due to the induced phase shift. Part of the light will pass through the polarization device, or analyzer, depending on the position of the light polarization vector at the output of the crystal. When the phase shift equals π, the light intensity passing through the modulator will be maximal. Consequently, the lower values of the phase shift will cause lower intensities at the output.

To optimize the modulator operation and ensure considerably large phase shift, an element performing some constant phase shift can be included in the modulation device. That element is, for example, an optical quarter-wave plate. At the same time, a Nikol prism, or a polaroid, can be used as a polarizer or an analyzer.

Amplitude modulators are often realized on a GaAs structure, because it possesses high thermal conductivity and good mechanical characteristics. Such a quality is very convenient for production of other optical devices, so integrated opto-electronic realization of several functional elements is possible. Beside GaAs, another very suitable material is $LiNbO_3$.

3.2.3 Phase Modulation of Optical Signals

The electro-optical crystal, such as that in Figure 3.1, is the basic element of an optical phase modulator. The input wave with plane polarization can propagate through the crystal only as an extraordinary ray, only if the optical axis of the crystal is properly positioned in relation to the propagation direction and the wave polarization plane. The phase velocity of the extraordinary ray can be changed under influence of the external electric field. Thus, the refractive index in the crystal is changed with the variations of the external electric field, leading to a change in the wave velocity in the crystal and to the induction of a wave-phase shift. We will consider that process using the equation of the monochromatic electromagnetic wave in the form

$$A(t) = A_m \cos(\omega t - \mathbf{k}\mathbf{s}x) \tag{3.29}$$

where A_m is the electrical vector amplitude, ω is the angular frequency, $\mathbf{s}$ is the unit vector in the x direction, and $\mathbf{k}$ is the wave vector with the intensity $k = 2\pi/\lambda$. The module of the wave vector in the crystal, having the refractive index n, is $k = 2\pi n/\lambda$. Hence, the initial phase of the wave is changed due to the changes in the refractive index in the crystal, and the phase modulation is imposed. It is important to know the value of the voltage needed to achieve the prescribed phase shift. That voltage can be evaluated using a procedure similar to that described for amplitude modulation. For example, the phase shift of π radians can be induced in crystals from the $\overline{4}2m$ group for the voltage [1, 2]

$$U = \frac{d}{\ell} \frac{\lambda}{n_0^3 r_{63}} \tag{3.30}$$

where d is the electro-optical crystal width, measured along the vector of the electric field, and ℓ is the crystal length measured along the direction of the wave propagation.

To minimize the voltage, U, which is necessary to cause the defined phase shift, the phase modulator can be designed as a multicrystal device. In such a way, the total length, ℓ, is increased; hence, a considerably smaller external voltage will cause the prescribed phase shift.

The mentioned phase modulators have a relatively narrow bandwidth. That bandwidth is characterized by the modulation signal bandwidth and should be as wide as possible. The spreading of the modulation bandwidth can be made by using a traveling electromagnetic wave as the external running field. In such a modulator, the broadband-phase modulation is performed due to mutual interaction between the modulating and the modulated signals in traveling wave–type systems. Thus, it is possible to reach the modulation bandwidth of several GHz at the relatively low power of the modulating signal.

It is well known that electromagnetic oscillations never have the ideal monochromatic nature, but they consist of several oscillations with different frequencies. Such oscillations form so-called *wave groups*, propagating in a real medium with a common, so-called *group velocity* [1]. In general, the group velocity differs from the corresponding phase velocity and has the form

$$v_g = \frac{d\omega}{dk} = v - \lambda\frac{dv}{d\lambda} = \frac{v}{1 - \dfrac{\omega}{v}\dfrac{dv}{d\omega}} \tag{3.31}$$

where v is the phase velocity of the wave, and ω is the angular frequency. The group velocity is equal to the phase velocity in a medium where there is no dispersion, or where $dv/d\omega = dv/d\lambda = 0$.

To realize broadband phase modulation in a traveling-wave modulator, electro-optical or magneto-optical material–based waveguides are used. These materials should satisfy the condition that the phase and group velocities of the propagating wave are equal in the wide region of frequencies. Such a condition can be imposed on a line with tranversal electromagnetic (TEM) type wave, filled particularly with electro-optical or magneto-optical material and particularly with dielectric material for an adjustment. Thus, the quantity of the electro-optical material with relatively high losses is minimized. It is important that the electro-optical and dielectric material have equal dielectric constants.

In most cases, the velocities of the light and the radiowave (or modulating signal) are different in the dielectric, which is caused by different values of the dielectric permittivity for different frequency regions. The equalization of the phase velocities can be imposed by inserting both a nonlinear material and a low-loss material into the conducting line, as illustrated in Figure 3.2. The synchronization condition is provided by prolonged mutual interaction between the radiowave and the optical wave. Two cases are possible in the propagation process: (1) when the radiowave and the optical wave are propagating in the same direction (collinear case), and (2) when they propagate in different directions (noncollinear case). The synchronization condition ($v = v_g$) in the collinear case is imposed by proper design of the conducting line, consisting of a waveguide line and crystal. In the noncollinear case, which is illustrated in Figure 3.2, the synchronization condition ($v = v_g \cos\varphi$) is imposed by the choice of the light coupling angle φ.

The role of the adjusting dielectric in the traveling-wave modulators can play the air, as well, and further analysis can be made under such an assumption. The phase modulation is realized by modulation of the propagating velocity of the light in the crystal. When the crystal length equals ℓ, the signal delay in the crystal is $\tau = \ell/v$, and the phase shift between the input and the output lightwaves is given as

$$d\Phi = \omega\tau = \omega\ell/v \tag{3.32}$$

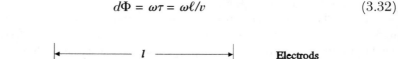

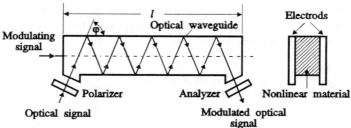

Figure 3.2 Scheme of the broadband phase modulator with traveling wave.

The phase shift $\Delta\Phi$ of the light signal depends on the deviation of the light velocity during its propagation through the crystal and is given as

$$\Delta\Phi = -\omega\ell\Delta v/v^2 \tag{3.33}$$

The light velocity in an electro-optical medium depends on the external acting electric field. If the light propagates in the y direction, and the electric field vector takes the x-axis position, the following relation is valid:

$$\frac{\Delta v_x}{c} = -\frac{r_{63}nE_z}{2} \tag{3.34}$$

where E_z is the z-coordinate component of the external modulating electric field. The coefficient r_{63} has the same meaning as in the amplitude modulator case (for the same type of crystal). The dependence of the phase shift $\Delta\Phi$ on the modulating external electric field can be found from (3.33) and (3.34) and has the form

$$\Delta\Phi = \frac{n^3\omega r_{63}\ell E_z}{2c} \tag{3.35}$$

Equation (3.35) shows that there is a linear dependence of the phase shift on the electric field of the external modulating signal. At the same time, the phase shift is also linearly dependent on the electro-optical crystal length. Thus, the crystal length is a parameter for adjustment of the phase-modulator characteristics.

3.2.4 Frequency Modulation of Optical Signals

3.2.4.1 Modulation by Use of the Zeeman Effect

The *Zeeman effect* refers to the phenomenon that the spectral line of the light source is split and shifted under the influence of a strong external magnetic field [1]. The normal Zeeman effect is characterized by the spectral split

$$\Delta\lambda = \mp \frac{1}{4\pi c^2}\frac{q}{m}\lambda^2 H \tag{3.36}$$

where q is the electron charge, m is the electron mass, and H is the intensity of the external magnetic field.

Modulation by use of the Zeeman effect is attended by some practical difficulties. A rather strong magnetic field is necessary to impose the broadband modulation, and complex additional devices are needed. Besides, broadband modulation is limited by a relatively high Q-coefficient of the optical resonator in the laser. The frequency modulation can be imposed by a homogeneous magnetic field for low modulation frequencies, while the high modulation frequencies need an inhomogeneous magnetic field.

3.2.4.2 Modulation by Use of the Stark Effect

The *Stark effect* has the same nature as the Zeeman effect, but it is induced by an electric field, rather than a magnetic field. The split of the spectral line and the frequency shift are proportional to the intensity of the electric field.

The linear Stark effect is characterized by linear dependence of the frequency shift on the intensity of an external electric field, but such an effect occurs only in hydrogen atoms. In all other atomic systems, the square-law Stark effect occurs, when the frequency shift is proportional to the square of the electric field intensity [1]. The square-law Stark effect is always negative, since the energy shift happens toward the low energy levels. The frequency modulators based on the Stark effect have imperfections similar to those based on the Zeeman effect, namely, the broadband frequency modulation needs a strong external modulating electric field, so the modulator becomes clumsy.

3.2.4.3 Modulation with Resonator Length Alteration

The change of the active optical length of a laser resonator is a relatively efficient way for a frequency modulation of the optical signal. The alteration of the optical length can be imposed if an electro-optical crystal is inserted into the resonator. The refractive index in the crystal is changed under the influence of external modulating electric field, so the optical length of the resonator is changed, causing the modification of the carrier frequency of the radiated optical signal.

The frequency shift caused by change in the optical length of the modulator is determined by the induced refractive index shift, Δn, by the geometrical length, ℓ_1 of the resonator, and by the length, ℓ_2, of the modulating crystal, so the following relation is valid:

$$\Delta f = \frac{c\ell_2 \Delta n}{\lambda \ell_1} \tag{3.37}$$

Another way to achieve the frequency modulation by changing the resonator optical length is related to a change in the geometrical length and can be imposed by mechanical action of the external electrical or magnetic field. The action is directed to the bottom facet of the resonator, which changes its relative position under the influence of the piezoelectric element. The relative shift of the facet means the simultaneous change of the optical-resonator length. The change, $\Delta\ell$, of the optical-resonator length causes the frequency shift of the generated optical signal, determined by the relation

$$\Delta f = \frac{f_0 \Delta \ell}{\ell} \tag{3.38}$$

where f_0 denotes the frequency of the optical signal before the action of the external modulating field.

3.2.5 Methods of Internal Modulation

All the aforementioned methods for amplitude, phase, or frequency modulation may be either internal or external depending on the modulator design. External modulation methods, in general, need more complex and more clumsy constituent elements, so internal modulation methods are applied whenever possible. This will not be the general rule in the future, however, because the design of the key elements will influence the choice of the modulation method.

Internal modulation methods are characterized by modulation of the optical wave parameters still in the lightwave-generation process. Internal modulation is based on the change in the laser active medium, laser resonator, or laser pump. All the modulation types discussed thus far can be imposed by internal modulation methods.

Internal modulation may be either synchronous or asynchronous, depending on the relation between the modulating signal frequency, ω, and the frequency shift, $\Omega = \pi c/\ell$, between the neighboring resonator longitudinal modes (c is light velocity in free space). Asynchronous modulation is characterized by the relation $\omega \ll \Omega$, while $\omega \simeq \Omega$ for synchronous modulation. Asynchronous internal modulation includes the amplitude modulation imposed by the laser pump current and the amplitude modulation imposed by the intrinsic resonator loss change. The change of intrinsic resonator loss is made by inserting the electro-optical crystal and the analyzer into the resonator and by the influence of the modulating electric field on such a structure. At the same time, frequency-modulation methods based on the change of the optical-resonator length and methods based on the Zeeman and Stark effects also can be included in internal modulation methods. The main imperfection of the asynchronous internal modulation method is its narrow band characteristics, caused by the large inertia of the restauration of new oscillations.

Synchronous internal modulation is imposed by change of the refractive index of the material in the resonator or by the change of the intrinsic resonator loss. The change of the internal refractive index is the most important practical method of frequency modulation in semiconductor lasers. This method is commonly used for modulation of optical-signal frequency in distributed feedback (DFB) lasers. The physical mechanism of the internal refractive index changes through changes in the density of free carrier in the semiconductor structure is explained in Section 3.3.4.2.

This review of modulation methods shows that there are many possibilities for modulator optimizations. The optimization is related to the choice of the most suitable materials for the resonator elements and to the design of the optical resonator. The optimization procedure must lead to the spread of modulation bandwidth, decrease of the modulation signal power, decrease of the optical power loss in the resonator, and so on. The most efficient realizations of optical modulators can be obtained by use of integrated optoelectronic devices. Some of these modulator realizations will be described at the end of this chapter.

3.3 SEMICONDUCTOR LASERS AS LIGHT SOURCES IN COHERENT OPTICAL TRANSMITTERS

3.3.1 Characteristics of Semiconductor Lasers

It is well known that semiconductor lasers are the most suitable sources in optical transmitters used in lightwave IM/DD systems [3]. There are several reasons for that, and all of them are related to the comparative advantages of laser diodes over other types of light sources. These comparative advantages include the relatively high output optical power, the small divergency angle of the output radiation, the relative narrow spectral line, and considerably small geometry.

But other requirements should be satisfied by the laser diodes in order to be applied in coherent optical transmitters. The laser diode must possess the high spectral purity (almost monochromatic operation), and high stability of the output wave parameters (phase and frequency stability above all). Although these requirements can be satisfied, maybe in a larger extent by other types of lasers, such as Nd:YaG laser, the small geometry and modulation simplicity are the most important in favor of laser diodes for coherent transmitters.

The width of the laser signal spectral line and the stability of the signal phase and frequency are determined by the presence of noise, disturbing the light signal coherence. This noise is often called the *phase noise*, and appears as a consequence of the spontaneous emission of the light in the laser. It is important to give the proper mathematical model of the laser noise, in order to develop methods for its suppression. The next subsection presents the main results related to laser noise modeling, using results published in papers [4–9] and results related to the mathematical description of the parameters of an output optical signal [9–12]. The most attention will be devoted to the results that are verified experimentally.

3.3.2 Noise Sources in Semiconductor Lasers

3.3.2.1 Noise Arising in the Stimulated Emission Process

According to quantum mechanical principles, electrons can populate only the discrete levels in the potential field of an atom [13]. An electron can cross from a lower to a higher energy level only by absorption of one photon with energy at least equal to the difference between the energy levels. At the same time, the transition from upper to lower energy level is followed by the emission of a photon. Such a transition may have either spontaneous or stimulated nature. Spontaneous transitions arise without any external acting, and the thermal-balance condition is the only cause of spontaneous transitions in an electron macro system. Stimulated transitions are caused by an external electromagnetic field. The electron macro system, consisting of a large number of micro systems, must be in the steady state.

In fact, the sum of spontaneous and stimulated transitions from the upper energy level to the lower energy level must be equal to the number of stimulated transitions from the lower to the upper energy level. Hence, if the lower energy level is denoted by the numeral 1 and the upper energy level by the numeral 2, the following well-known equilibrium equation can be established [1]:

$$\sigma(f_{12})B_{21}N_1 = A_{12}N_2 + \sigma(f_{12})B_{21}N_2 \tag{3.39}$$

where N_1 is the number of micro systems (actually, electrons) in a state 1, N_2 is the number of micro systems in a state 2, A_{12} is the Einstein coefficient for spontaneous emission, $B_{21} = B_{12}$ is the Einstein coefficient for stimulated emission, and $\sigma(f)$ is the space density of the external electromagnetic field evaluated in a unity interval around the frequency f.

The coefficient A_{12} is a measure for the mean number of spontaneous transitions per second for every micro system in the state 2. The reciprocal value of this coefficient, denoted by τ_{12}, is the average lifetime of a micro system in state 2. The lifetime is finished only by a spontaneous transition to state 1. The parameter τ_{12} has exact meaning only in a system without any external field influence. Besides, the parameter τ_{12} depends on the other possible transitions in the macro system.

The Einstein coefficient A_{12} can be determined from the thermal-balance condition in a macro system, using the Boltzmann distribution for the energy-level population [13]:

$$\frac{N_1}{N_2} = \exp\left(-\frac{W_2 - W_1}{k\Theta}\right) = \exp\left(-\frac{hf_{12}}{k\Theta}\right) \tag{3.40}$$

where W_1 and W_2 are the energies of corresponding levels, k is Boltzmann's constant, and Θ denotes the absolute temperature. Equation (3.40) describes the distribution that occurs with the largest probability in a macro system, based on the large number of independent micro systems.

According to (3.39), the number of micro systems on a defined energy level is reciprocally proportional to the total level energy. The decrease in the number of micro systems is described by falling exponential law. From (3.39) and (3.40), the next relation between the Einstein coefficients can be obtained:

$$A_{12} = B_{12}\sigma(f_{12})\left[\exp\left(\frac{hf_{12}}{k\Theta}\right) - 1\right] \tag{3.41}$$

Since the coefficient A_{12}, which describes the probability of a random process, must be independent of temperature, it must be

$$\sigma(f_{12}) = \frac{K(f_{12})}{\exp\left(\dfrac{hf_{12}}{k\Theta}\right) - 1} \tag{3.42}$$

The unknown function, $K(f)$, can be determined as a product of an individual oscillation energy and the oscillation density per frequency unit (see Appendix I), so we have

$$K(f) = \frac{8\pi hf^3}{c^3} \tag{3.43}$$

Now (3.42) takes the form of the well-known Planck's distribution:

$$\sigma(f) = \frac{8\pi hf^3}{c^3\left[\exp\left(\dfrac{hf}{k\Theta}\right) - 1\right]} \tag{3.44}$$

The Planck's relation, in fact, describes the radiative noise generated in a passive device. This noise component is caused by spontaneous emission from the upper level and has a thermal nature. The thermal noise will vanish if the absolute temperature reaches the zero value. But an active device, such as a laser, is characterized by inversion population, or so-called *inversion temperature* when $N_2 > N_1$. Thus, the noise appears even in the case when the temperature is zero. Such a noise, known as the *quantum one*, is caused by the laser pump and will be the subject of the next consideration.

The inverse population is characterized by the photon density parameter:

$$n = \frac{N_2}{N_1 - N_2} = \frac{1}{\left[\exp\left(\frac{hf_{12}}{k\Theta}\right) - 1 \right]} \tag{3.45}$$

The inverse population is the precondition for stimulated emission and light-signal amplification. Thus, the monomode electromagnetic wave will be amplified by being passed through the medium with imposed inverse population. This process can be described by the equation for the radiated photon density n, or

$$\frac{dn}{dt} = A_{12}N_2 + \sigma(f_{12})B_{21}N_2 - \sigma(f_{12})B_{21}N_1 \tag{3.46}$$

$$= A_{12}n(N_2 - N_1) + A_{12}N_2$$

The photon density at the output end of the structure, at the moment $t = t_0$, can be found from (3.46) for an initial condition $n = n_0$ (at the moment $t = 0$), that is,

$$n = n_0 G + \frac{N_2}{N_2 - N_1}(G - 1) \tag{3.47}$$

where G is a gain coefficient of the optical signal in the laser structure, given as

$$G = \exp[A_{12}(N_2 - N_1)t_0] \tag{3.48}$$

Thus, it can be seen that the photon density at the output is nearly equal to the product of the input photon density and the gain coefficient but corrected by the product of the gain, G, and the term

$$n_{sp} = \frac{N_2}{N_2 - N_1} \tag{3.49}$$

The number n_{sp}, in fact, defines the photon number density of the quantum noise due to spontaneous emission. The total energy of the quantum noise per mode is determined by the product of the noise photon density and the energy of one photon:

$$W_{nc} = \frac{N_2}{N_2 - N_1}hf \tag{3.50}$$

The total power of the quantum noise can be evaluated as the product of the energy, W_{nc}, and the laser frequency bandwidth Δf, that is,

$$P_{nc} = W_{nc}\Delta f \tag{3.51}$$

Since the total noise power consists of the quantum and thermal noise powers, it is necessary to add the thermal component to the quantum component defined by (3.51). The power of the thermal noise can be obtained as the product of the thermal noise energy, W_{nt}, and the frequency bandwidth, Δf. The thermal noise energy is, in fact, the ratio of the value defined by (3.44) and the total number of modes. Because the total number of modes (or individual oscillations) is n' ($n' = 8\pi f^2/c^3$) (see Appendix I), the power of the thermal noise becomes

$$P_{nt} = \frac{hf\Delta f}{\exp\left(\dfrac{hf}{k\Theta}\right) - 1} \tag{3.52}$$

Thus, the total power of the noise in the semiconductor laser can be written as

$$P_n = hf\left(\frac{N_2}{N_2 - N_1} + \frac{1}{\exp\left(\dfrac{hf}{k\Theta}\right) - 1}\right)\Delta f \tag{3.53}$$

3.3.2.2 Laser Noise Spectra

Since the laser noise is the random process, analysis of it is done by standard methods, using the autocorrelation function, the cross-correlation function, and power spectral density. It is interesting to consider those functions for the noise amplitude and the noise frequency, as well as for the spectral function of the electrical field of the generated optical signal. The electric field phasor, $E(t)$, of the generated light signal can be expressed by the signal and noise components, or as

$$E(t) = [A_0 + a_n(t)]\exp\{j[\varphi_0 + \varphi_n(t)]\} \tag{3.54}$$

where A_0 and φ_0 are the stationary values of the amplitude and the phase, respectively, while $a_n(t)$ and $\varphi_n(t)$ are the fluctuations of those values, determining the total noise.

The amplitude spectrum, $S_A(f)$, is defined as the power spectral density of the amplitude fluctuations, $a_n(t)$, or

$$S_A(f) = \lim_{T\to\infty} \frac{1}{T} <A_n^*(f)A_n(f)> \tag{3.55}$$

where $A_n(f)$ is the Fourier transform of the function $a_n(t)$ on time interval from $-T/2$ to $T/2$. The angle brackets ($<>$) mean the average value over an assemblage, while the asterisk denotes the conjugate-complex value.

The spectrum of the frequency noise, $S_F(f)$, is defined as the power spectral density of the instantaneous frequency fluctuation of the generated light signal. The function of the instantaneous frequency fluctuations is defined as

$$f_n(t) = \frac{1}{2\pi} \frac{d\varphi_n}{dt} \tag{3.56}$$

so, the corresponding power spectral density is

$$S_F(f) = \lim_{T\to\infty} \frac{1}{T} <F_n^*(f)F_n(f)> \tag{3.57}$$

The function $F_n(f)$ is the Fourier transform of the function $f_n(t)$.

The cross-correlation spectrum, $S_{AF}(f)$, between AM and FM noises is defined as

$$S_{AF}(f) = \lim_{T\to\infty} \frac{1}{T} <F_n^*(f)A_n(f)> \tag{3.58}$$

Instead of the function $S_{AF}(f)$, the so-called coherency coefficient, k, and the phase angle θ are used more often. These parameters are defined, respectively, as

$$k^2 = \frac{|S_{AF}(f)|^2}{S_A(f)S_F(f)} \tag{3.59}$$

and

$$\theta = \arctan[S_{AF}(f)] \tag{3.60}$$

The electric field spectrum of the generated light signal is defined as the power spectral density function of the electric field phasor. Since the amplitude noise has considerably smaller value compared with the signal amplitude, the component $a_n(t)$ in (3.54) can be neglected. Thus, the spectral function of the electric field can be evaluated by the Fourier transform of the electric field autocorrelation function. The normalized autocorrelation function of the electric field is defined as

$$R(\tau) = \frac{<E(t)*E(t + \tau)>}{A_0^2} \tag{3.61}$$

From (3.54), (3.56), and (3.61), the next equation follows:

$$R(\tau) = <\exp j\Delta\varphi_n(\tau)> \tag{3.62}$$

where

$$\Delta\varphi_n(\tau) = \varphi_n(t + \tau) - \varphi_n(t) = 2\pi\int_t^{t+\tau} f_n(z)dz \tag{3.63}$$

It is suitable to express the phase difference, $\Delta\varphi_n$, through the phase variance, σ_φ^2, by using the standard methods applied in random-process analysis. Thus, for a standard zero-mean Gaussian process, we have [14]

$$R(\tau) = \exp(-\sigma_\varphi^2/2) \tag{3.64}$$

The phase noise variance, σ_φ (denoted in Chapter 2 by σ_n) is determined by

$$\sigma_\varphi^2 = <\Delta\varphi_n(\tau)^2> = 4\pi^2\int_t^{t+\tau}\int_t^{t+\tau} F(t' - t'')dt'dt'' \tag{3.65}$$

Since the autocorrelation function of the function $f_n(t)$, which defines instantaneous frequency deviations, is given as

$$F(t - t') = <f_n(t)f_n(t')> \tag{3.66}$$

(3.65) can be easily transformed into the form

$$\sigma_\varphi^2 = 8\pi^2\int_0^\tau(\tau - t)F(t)dt \tag{3.67}$$

Now, keeping in mind the Fourier relation

$$F(t) = \int_0^\infty S_F(t)\cos 2\pi ft \, df \tag{3.68}$$

the functional dependence between the phase noise variance and the spectral function of the FM noise can be obtained. Thus, from (3.67) and (3.68), it follows that

$$\sigma_\varphi^2 = 4\int_0^\infty S_F(f)\sin^2(\pi f\tau/f)df \tag{3.69}$$

The phase-noise variance is the most important parameter of the phase noise, determining the characteristics of the coherent transmission systems, as was shown in Chapter 2.

The spectral function $S(f)$ of the electric field is found by the Fourier transform of (3.62) or (3.64), with the value σ_φ^2 defined by (3.69). The function $S(f)$ is widely used as a measure of the laser radiation time coherency.

Thus, we have introduced all the parameters that characterize laser noise. The next step in the theoretical consideration is the evaluation of these parameters. Several approaches for such evaluation have already been presented, and have agreed with experimentally measured values. Theoretical models that consider laser noise characteristics have been given in the literature [4–7, 11]; we will present and discuss only the main results obtained in the mentioned models, which will be enough for the next systems considerations.

3.3.2.3 Theoretical Analysis of Laser Noise

The theoretical models of the noise in a semiconductor laser presented so far have been based mainly on quantum mechanical principles. The laser-noise parameters are obtained from established equations that describe the so-called *noise force*. Such equations are Langevin equation, density matrix equation, Fokker-Planck equation, circuit theory equation, and so on. The Langevin equation [5] is used most often because it gives a clear physical representation of the noise generation and leads to the results that agree with experimental ones. Several approaches are based on the Langevin equation [6, 7, 10, 15], with nearly equal results for the main noise parameters, which we will point out.

The spectral densities of the AM and FM noises in the first approximation can be expressed by the constant values

$$S_A(f) = \frac{A_0^2 \delta f'}{\pi} \frac{1}{1 + (\gamma_s/2\pi)^2} \tag{3.70}$$

$$S_F(f) = \frac{(1 + \alpha^2)\delta f'}{\pi} \tag{3.71}$$

where

$$\delta f' = \frac{\pi n_{sp} h f \Delta f}{A_0^2} \tag{3.72}$$

Parameters n_{sp} and A_0 are the parameters of spontaneous emission and the amplitude of the electric field, defined by (3.49) and (3.54), respectively; Δf is the frequency bandwidth of the laser resonator; γ_s is the so-called *depletion constant* [see (3.81)]; and α is the so-called *correction term* of the spectral linewidth. The correction term is determined by the ratio

$$\alpha = \frac{\Delta n'}{\Delta n''} \tag{3.73}$$

where $\Delta n'$ and $\Delta n''$ define the changes in the real and imaginary parts, respectively, of the refractive index under the pump influence. This correction factor is the subject of theoretical and experimental investigations [9, 10]. Thus, the main conclusion is that, in the first approximation, both the AM and the FM noises have the nature of white noise with the magnitude depending on the output optical power.

The coherency coefficient, k, and the phase angle, according to (3.59) and (3.60), can be evaluated as

$$k^2 = \frac{\alpha^2}{1 + \alpha^2} \tag{3.74}$$

and

$$\theta = \begin{cases} 0 \text{ for } \dfrac{\partial n'}{\partial n} > 0 \\[2mm] \pi \text{ for } \dfrac{\partial n'}{\partial n} < 0 \end{cases} \tag{3.75}$$

The power spectrum of the electric field is given as a Lorentz function:

$$S(f) = \frac{\partial f'(1 + \alpha^2)}{2\pi\{f^2 + [\delta f'(1 + \alpha^2)/2]^2\}} \tag{3.76}$$

with the spectral linewidth of the output optical signal:

$$\delta f = (1 + \alpha^2)\delta f' \tag{3.77}$$

The relatively simple equations presented thus far are suitable for consideration of coherent system characteristics but are valid only for frequencies far enough from some resonant frequency. More exact solutions can be obtained by taking into account the resonant effect in the semiconductor lasers. The evaluation of exact solutions is rather complex, so we will write only the final expressions [16]:

$$S_A(f) = \frac{A_0^2 \delta f'}{\pi} \frac{f^2 + (\gamma_s/2\pi)^2}{(f^2 - f_R^2)^2 + (\gamma_s/2\pi)^2 f^2} \tag{3.78}$$

$$S_F(f) = \frac{\delta f'}{\pi} \left[1 + \frac{\alpha^2 f_R^4}{(f^2 - f_R^2)^2 + (\gamma_s/2\pi)^2 f^2} \right] \tag{3.79}$$

where f_R and γ_s are the resonant frequency and the early mentioned depletion constant, respectively. These parameters are given as

$$f_R^2 = \frac{A_0^2 G(n_0)}{(2\pi)^2} \frac{\partial G}{\partial n} \tag{3.80}$$

$$\gamma_s = k + A_0^2 \frac{\partial G}{\partial n} \tag{3.81}$$

The initial photon density number, n_0, the photon density number, n, and the optical gain, G, were all introduced in (3.47). The typical shapes of the spectral functions of the AM and FM noises, in accordance with (3.78) and (3.79), are shown in Figure 3.3. The functions are normalized with the values $S_A(0)$ and $S_F(0)$, respectively. The resonant frequency value is taken to be about 1 GHz, according to the experimental results [9–11]. The peak values of the functions from Figure 3.3 are moved toward the higher frequencies for an increase of the output optical power.

It is important to point out the main conclusions once again. First, the noise spectrum functions are practically constant, not only before the resonant frequency, but after that frequency, as well. Second, the resonant frequency is directly proportional to the output optical power.

There is another component of laser noise in the FM spectrum. This component is often called $1/f$ noise, since its influence corresponds to the term K/f [17] (K is a constant). The term K/f is to be added on the right side of (3.79). The constant K is strongly dependent on the output optical power, so the influence of the $1/f$ term increases with the higher levels of the output laser signal.

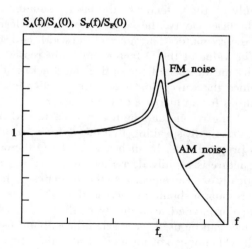

Figure 3.3 The illustration of the power spectrum functions for AM noise and FM noise.

Finally, it is useful to know the relative values of the parameters of laser-noise characteristics, especially for the spectral width and the correction parameter, α. These values were obtained experimentally [10, 16, 18]. Thus, the product of the output power and the spectral width $(\delta f P_0)$ for GaAlAs lasers is in the range 20 to 114 MHz mW, while for InGaAsP lasers that product is in the range 15 to 200 MHz mW. At the same time, the value of the correction coefficient, α, is in the range 2.2 to 6.2 for GaAlAs lasers and from 2.2 to 6.6 for InGaAsP lasers.

3.3.3 Stabilization of Laser Characteristics

The following main requirements should be satisfied by a semiconductor laser in a coherent lightwave communications system: a narrow spectral linewidth (below 0.1 MHz), high stability of the carrier optical frequency, the possibility of phase tracking, and the possibility of direct modulation of the carrying wave. All these characteristics, except the last one, are important both for the optical source in the coherent transmitter and for the local optical generator in the optical receiver. Some modulation schemes include the local optical generator even in the optical transmitter, to provide proper stabilization of the output optical-signal characteristics.

3.3.3.1 Narrowing of the Output-Signal Spectral Linewidth

The spectral linewidth of the output optical signal of ordinary semiconductor lasers is determined mainly by the Q factor of the laser resonator. This factor is defined by the length of the laser cavity, the value of the optical feedback power, and the waveguide loss in the resonator. The spectral linewidth is decreased by an increase of the factor Q. The value of the Q factor can be increased by an increase of the resonator length, by an increase of the optical feedback, and by a decrease of the waveguide loss. Since the waveguide losses are practically constant, there are, in effect, two possibilities for an increase of the Q factor.

An increase of the resonator length can be achieved by the insertion of an external resonator. Practically speaking, that means inclusion of an additional resonator structure with proper facets. In such a way, the Q factor is increased, but the stability of the structure is rapidly decreased. Increase of optical feedback can be obtained either optically or electronically. In the former case, the reflection coefficients of the facets in the resonator should be increased, while in the latter case, influence is exerted on the pump current and, accordingly, on the feedback-signal level. Integrated optoelectronic devices can be used for efficient broadband feedback.

Increase of the Q factor can be most efficiently made by insertion of the external facet and formation of an external resonator. In such a way, both resonator length and optical feedback are increased, leading to considerable decreasing of the spectral linewidth. There are two methods for increasing the Q factor: the external mode operation, as illustrated in Figure 3.4, and self-injection locking.

The external operation requires the decrease of the reflection coefficients of ordinary structure facets by coating them with an antireflection layer. Thus, the ordinary resonator loses that role, and the external resonator between two external facets becomes predominant. Now the laser structure plays the role of an optical amplifier. The resonant frequency of the new structure is determined by the external cavity. The diffraction grating performs the function of the mode selector and provides the monomode operation at a selected carrier wavelength. This operational principle is applied in the distributed feedback (DFB) lasers. Thus, if the length of a new resonator is about 10 cm, the spectral linewidth of the output optical signal will be about 10 KHz, instead of several MHz, as in an ordinary structure [18].

When the reflection coefficients are not decreased as required, the locking with the external resonator is weak, and the resonant frequency of the output signal is determined not only by the length of the external resonator but by the resonant frequency of the main laser cavity, as well. That state, known as a *self-injection locking* [19], can be imposed only by one external half-mirror that reflects back to the main resonator a relatively small amount of the energy (the greater part of the energy is going out as an output signal). The main parameter that determines the resonant frequency of the self-injection structure is the time delay, τ, necessary for one return traveling of the light through the external resonator. That delay is given as the ratio $2\ell/c$, where ℓ is the external resonator length, and c is the light velocity. Analysis of the self-injection locking structure operation is made in the same way as was done for the ordinary semiconductor laser structure [5–10].

The frequency, f, of the oscillation of a new extended structure can be expressed as the sum of the oscillation frequency, f_0, of the ordinary laser structure and a frequency deviation, Δf. The frequency deviation is determined numerically [19] and is proportional to the delay, τ, the correction factor, α, and the optical feedback coefficient. It is important to know that the oscillation frequency, f, is not stable

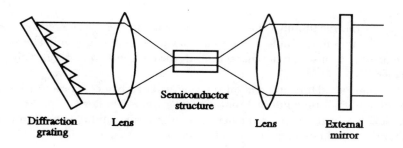

Figure 3.4 The increasing of the Q factor of semiconductor laser by an external operation.

and can be changed during the laser operation. The operation instability is directly proportional to the frequency deviation, Δf. On the other hand, the spectral linewidth of the output signal from a new structure is inversely proportional to the frequency deviation, Δf, so it is important to find an optimal deviation and provide the stable operation of the device. The stable operation also is important, since the new structure is sensitive to random deviations in the external resonator parameters, which can be caused, for example, by random variations of the external temperature or mechanical instability.

We will not discuss the application of the electric PLL for the narrowing of the spectral linewidth of the semiconductor lasers, because that method is, in essence, similar to the method of the stabilization of the optical carrier frequency, which is discussed next.

3.3.3.2 Stabilization of the Optical Carrier Frequency

Carrier frequency of the output laser signal fluctuates considerably around some average value. These fluctuations are transmitted at the receiving end and are included in the current signal in the IF band. That leads to decreased coherent optical receiver sensitivity and an increased BER. Besides that, the completely stabilized laser oscillator in the transmitter, or the so-called master oscillator, has the important role in the frequency stabilization of the whole transmission system, especially in the networks where the WDM technique is applied.

Laser operation stability may have both the short-term stability character and the long-term stability character. Frequency stability is determined by the ratio of the average frequency of the output optical signal and the frequency fluctuations range, $\Delta f(\tau)$, in the considered time period, τ, or

$$S = \frac{1}{\tau} \int_0^\tau \frac{f(t)dt}{(f_{max} - f_{min})} = \frac{\overline{f}}{\Delta f(\tau)} \qquad (3.82)$$

where f_{max} is the maximum value of the frequency in the considered time interval, and f_{min} is the minimum value of the frequency in the same interval. The stability of the frequency is also often characterized through the reciprocal value of the parameter S.

The considered time interval must be defined in advance. The short-term stability is evaluated in a time interval that is shorter than the allowed time, τ_0, necessary for the device to register the frequency. The long-term stability is not limited by the value τ_0 and is evaluated for the period $\tau > \tau_0$. The short-term stability is determined by the character of the noise in the laser-signal generation process, while the long-term stability is defined in relation to some rational time interval (minute, hour, day), when the variations of the parameters influencing the stability are expected. Such variations include temperature variations, and pump current variations.

It has been shown that the factors that determine the long-term stability of the semiconductor laser are, above all else, changes in the geometrical length, ℓ, of the resonator or the refractive index value, n, in the resonator. Thus, the long-term stability can be evaluated by

$$\frac{1}{S} = \frac{\Delta\ell(\tau)}{\ell} + \frac{\Delta n(\tau)}{n} \tag{3.83}$$

The meaning of *laser operation stability* is closely related to so-called *frequency reproducibility*, which defines the accuracy of the frequency reproduction after numerous on-off settings of the laser operations. The reproducibility is defined by the same relation as stability, but now Δf in (3.82) corresponds to standard deviations of the frequency in the on-off setting process or in the system adjustment process.

Different optical or electrical methods can be applied for stabilization of the laser carrier frequency. The regulative schemes are based on the principles of an optical or an electrical PLL. The regulation methods take part both in the optical transmitter, for the carrier frequency regulation, and in the optical receiver for the local oscillator adjustment. Adjustment of the local oscillator is made to provide proper tracking of the incoming optical carrier frequency. Herewith, the methods related to carrier frequency regulation in the optical transmitter will be described; the methods applied in an optical receiver will be analyzed in Chapter 4.

The central, or carrier, frequency of the semiconductor laser signal can be changed under the influence of the active layer temperature, because the refractive index of the active layer depends on the temperature. On the other hand, the changes of the carrier frequency can be induced by fluctuations in the pump current or by an aged effect. The carrier frequency variations under the influence of the temperature fluctuations are approximately in the range 10 to 20 GHz/K, while the same variations for the pump fluctuations are about 1 to 5 GHz/mA [20]. The curve of the frequency-temperature dependence does not have a flat character. There are flat parts that correspond to the narrower time intervals, with the slope 10 to 20 GHz/K and saw parts that correspond to the frequency jumps between the longitudinal modes in the laser. The slopes of these parts are above 100 GHz/K. The flat parts belong only to one mode and are related to the thermal dependence of the refractive index and the thermal spreading of the resonant cavity. The carrier frequency jumps to the value corresponding to the next longitudinal mode at the moment when the gain coefficient of that mode becomes large enough due to the maximum gain shift induced by temperature change. Frequency jumps rarely happen in DFB lasers, because there is the largest difference between the threshold gain levels of the longitudinal modes. That difference, determined by the frequency selectivity of the diffraction grating, is larger than induced thermal variation of the gain. Oscillation frequency in a longitudinal mode can be changed in the range of about 30 GHz. Thus, frequency adjustment and stabilization are possible in the same range. The widest regulation range is in DFB lasers [20].

Efficient stabilization of the carrier frequency in semiconductor lasers is mainly performed by two methods. The first method is automatic control of the frequency by the laser temperature control; the second is control by the injection current control. Frequency control by the temperature control provides the wider range of the frequency adjustment (nearly about 30 GHz), while frequency control by the injection current control provides a considerably narrower adjustment range. Control by the injection current provides the shortest time response at the same time. All the mentioned regulation methods need some standard as the regulation reference. Such a standard is the Fabry-Perot interferometer, but an absorption line of the gas molecules, such as NH_3, CO_2, and CH_3Cl_2, also can be used.

Synchronous detection also can be applied in the case when absorption coefficients of the gas molecules are too small. In such a case, the sine modulation of the laser radiation passing through the gas cell is applied, and then its detection is made. The part of the output electrical signal goes back to the input, and the classical electrical feedback is formed. The standard for the frequency stabilization is the point where the first derivative of the absorption curve is zero.

Frequency regulation is performed by a regulative scheme with either one feedback loop or a double feedback loop. The double feedback loop offers higher stability. In such a case, the optical signal from the laser is split into two parts. One part is directly detected by a photodiode, while the other passes through a Fabry-Perot interferometer (or some other standard) and then is detected by another photodiode. The attenuation of the optical signal is equalized by a tunable optical attenuator placed before the photodiode with direct detection. The stabilization is made at the resonant frequency, f_r, of the interferometer (f_r is equal to $mc/2n\ell$, where c is the light velocity, m is an integer, and ℓ is the resonator length). The current signals from the photodiodes go to the differential amplifier, and the output signal from the amplifier goes back to correct the injection current of the laser. Thus, the classical feedback loop is formed. Another loop can be formed by use of the Peltier element, as in standard laser-regulation schemes.

3.3.3.3 Characteristics of Optical Phase-Locked Loop

The optical phase-locked loop (OPLL) is widely used not only for regulation schemes in the optical transmitter but also for frequency adjustment of the local optical oscillator in different optical modulation-demodulation schemes. An optical locked loop circuit consists of the frequency-tuning laser, a photodiode, and the loop filter. The photodiode is used as a phase detector. Thus, there are the same elements as in the standard electric PLL which is examined in the next chapter. Because of that, we will describe the characteristics of the optical locked loop only briefly. The scheme of an optical locked loop is shown in Figure 3.5.

Optical signals from the referent laser and from the local laser oscillator are mixed before they reach the photodiode. The referent laser is often called the *master laser*, and the frequency adjusting laser is called the *slave laser*.

The electric field of the incoming master laser signal can be written in the form

$$E_m(t) = E_m \sin[2\pi f_m t + \varphi_m(t)] \tag{3.84}$$

while the electric field of the slave laser has the form

$$E_s(t) = E_s \sin[2\pi(f_s + \Delta f)t + \varphi_s(t)] \tag{3.85}$$

where f_s is the oscillation frequency in a free-running operation, and Δf is the frequency-deviation range due to optical feedback. The output voltage from the photodiode is given as

$$V_d = (K_d/2)\{E_m^2 + E_s^2 - E_m E_s \sin[2\pi(f_s - f_m + \Delta f)t - \varphi_m(t) + \varphi_s(t)]\} \tag{3.86}$$

where K_d is a photodetection constant. The direct voltage component is eliminated by being passed through the phase detector, and the rest of the voltage goes to the loop filter. The output voltage from the filter can be expressed as

$$V(t) = -K_f \int_0^t \sin \psi(\tau) h(t - \tau) d\tau \tag{3.87}$$

where $h(t)$ is the impulse response of the filter, and $\psi(t)$ is equal to

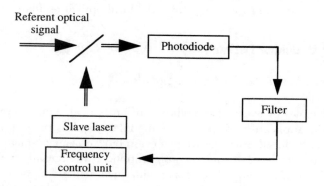

Referent optical
signal

Photodiode

Filter

Slave laser

Frequency
control unit

Figure 3.5 The scheme of the optical locked loop.

$$\psi(t) = 2\pi(f_s + \Delta f - f_m)t + \varphi_s(t) - \varphi_m(t) \tag{3.88}$$

The constant K_f now becomes

$$K_f = K_d E_m E_s / 2 \tag{3.89}$$

The voltage, $V(t)$, together with the frequency control unit, is used to control the output frequency of the slave laser in some fashion (for example, by varying the electrical drive current to the laser or by adjusting the laser cavity length).

$$\frac{d\varphi_s}{dt} + 2\pi\Delta f = K_s V(t) \tag{3.90}$$

where K_s is a proportionality constant. Keeping in mind (3.87), the following differential equation for the phase of the local optical oscillator can be written:

$$\frac{d\varphi_s}{dt} + 2\pi\Delta f = -K_s K_f \int_0^t \sin \psi(\tau) h(t - \tau) d\tau \tag{3.91}$$

This differential equation defines the characteristics of the OPLL. The form of the equation is, in fact, the same as the form of the characteristic equation of the electric PLL (see Chapter 4). Hence, the method of the solution of (3.90) is practically the same as that applied for the electric locked loop. Equation (3.90) has a general form valid not only for modulation process in the optical transmitter and for demodulation process in the optical receiver but also for carrier-frequency stabilization process in the laser. Stabilization of the carrier frequency has already been discussed, so it will be mentioned only that in such a case the optical signal passing through the standard should be taken instead of the signal from the local optical oscillator in (3.91).

3.3.4 Optical Modulator Design

3.3.4.1 External Modulation of Optical Signals

The basic principles of external modulation of an optical signal were described in Section 3.2. The application of an external modulator is illustrated in Figure 3.6(a). The modulator is placed after the frequency stabilized semiconductor laser, which is, in fact, the master laser in such a configuration. At the output of the modulator, the different types of modulated signals are obtained, depending on the chosen type of modulator and the applied modulation effect. It is most useful to use the modulation methods that provide the largest sensitivity of the coherent optical receiver, so the PSK and DPSK methods have the advantage. Because of that, we will describe

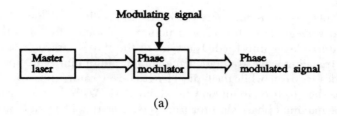

(a)

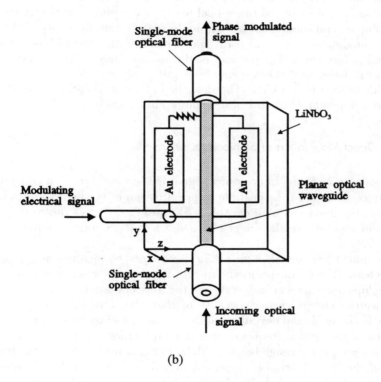

(b)

Figure 3.6 (a) The scheme of the coherent optical transmitter with the external optical modulator, and (b) the phase modulator base on LiNbO₃. (*After:* [21], © 1987 IEEE, reprinted with permission.)

the solution of the DPSK modulator realized in the planar waveguide technique. Undoubtedly, integrated optoelectronic versions will be the favorable solutions in the future. As an illustration, the functional scheme of a phase modulator made on the LiNbO₃ structure is shown in Figure 3.6(b).

The LiNbO$_3$ crystal is cut along the x-axis and is directed along the y-axis. The planar optical waveguide with such a structure is obtained by the diffusion of the 9-μm-wide titanic layer into LiNbO$_3$ crystal in the x-axis direction [21]. The electrodes of the traveling wave are made from gold or aluminum. The typical thickness of the electrodes are 3μm, while the typical lengths are 10 mm. Thus, the characteristic impedance of the electric circuit will be about 8 Ω. With such a structure of the electrodes, the maximal phase shift for transverse electric (TE) polarization (parallel to the z-axis) of the input signal is reached. The input and output ends of the waveguide have antireflection coatings, decreasing the losses at the coupling points between the optical fiber and the optical planar waveguide. The described structure is very compact and suitable for use within the optical transmitter. The internal losses of such a modulator are about 6 dB and depend on the optical signal wavelength and signal polarization. The product of the voltage, necessary to obtain the phase shift of π radians, and the waveguide length is about 7 Vcm, for a modulation bandwidth from 0 to 3.5 GHz. The described phase modulator shows the way for the future improvement of optical modulator characteristics.

3.3.4.2 Direct Modulation of Semiconductor Lasers

Direct modulation of the laser optical signal is commonly used in coherent optical transmitters, because of the possible compact design of the light source and optical modulator. By direct modulation, ASK, FSK, and PSK signals can be obtained. The modulation schemes for the direct phase and frequency modulations are shown in Figure 3.7.

An optical FSK signal (Figure 3.7a) is obtained by direct current injection into the slave laser. The frequency modulation of the slave laser is reached in a few steps. First, the injection current changes the temperature of the active layer and the density of the electron charge carrier. Because of that, the refractive index in the active structure is changed, and deviations of the output signal frequency are induced. The deviations of the optical frequency due to temperature effect are dominant for the modulation frequency range below 1 MHz. For this frequency bandwidth, the frequency of the optical signal is decreased with the injection current increase; hence, the modulated signal and the modulating current have the opposite phases. Above the modulation frequency of 10 MHz, the optical signal frequency deviations are caused by the injection current effect, which can be explained by the PLL model.

The slave laser is, in fact, an FM laser, but it is not enough for efficient frequency modulation. Because of that, an electric feedback, as in Figure 3.7(a) should be applied. In such a design, the part of the FM optical signal is mixed with the signal of the local master laser, and heterodyne detection is performed. The electric signal from the photodiode passes through the frequency discriminator and the loop filter and modulates the injection current of the slave laser. It should be noted that the demodulated electric signal changed the phase; thus, the negative feedback is realized.

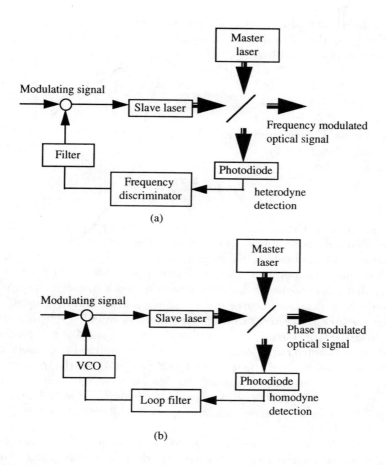

Figure 3.7 Coherent optical transmitters for direct frequency and direct phase modulation: (a) frequency modulator, and (b) phase modulator.

The negative feedback suppresses the FM noise and nonuniform frequency response and contributes to improvement of the frequency modulator characteristics.

The direct phase modulation illustrated in Figure 3.7(b) is made by current injection into the slave laser. The phase behavior of the output signal from the slave laser can be described by the PLL model. When direct modulation of the slave laser is made by the time signal, $g(t)$, (3.90), which describes the phase φ behavior, takes the form

$$\frac{d\varphi}{dt} = -K\varphi + 2\pi g(t) \tag{3.92}$$

where it is assumed that the phase of the master laser is not modulated ($\varphi_m(t) = 0$), so that $f_s = f_m$. At the same time, it is assumed that $h(t) = \delta(t)$ (Dirack pulse). The Laplace transformation of (3.92) leads to the expression

$$\Phi = \frac{2\pi G(s)}{s + K} \tag{3.93}$$

where $G(s)$ is the Laplace transform of the function $g(t)$. When it is $|s| \ll K_s K_f = K$, the following relation is valid:

$$\Phi = \frac{2\pi}{K} G(s) \tag{3.94}$$

It can be seen that the phase Φ is modulated by the signal $g(t)$, which means that conversion of the frequency modulation to the phase modulation is made. The upper value of the modulation frequency is determined by the frequency bandwidth of the PLL. Since the loop bandwidth is determined by the signal delay in the loop, it is clear that the decrease of delay increases the upper modulation frequency. To decrease the loop delay, integrated optoelectronic schemes must be used.

The essential characteristics of the described modulation schemes are the flat FM response, the large FM efficiency, and suppression of parasite amplitude modulation. All these characteristics can be achieved by using modified lasers with DFB [22], where the modulation efficiency of 1 to 2 GHz/mA and the upper modulation frequency above a few hundred megahertz can be achieved. The lasers with DFB can be modified in different ways, such as by multielectrode modulation. In such a way, the flat amplitude response and the upper modulation frequency over 1 GHz can be achieved [23, 24].

It seems that so-called multiple quantum well (MQW) DFB lasers can offer many advantages compared with other semiconductor laser realizations. These lasers have a significantly lower linewidth floor (down to 600 kHz) than other realizations [25], and a reduced noise figure, which leads to significant SNR improvement [26].

Direct amplitude modulation can be performed the same way as phase modulation. The main problem is in the suppression of the phase and frequency deviations when the injection current is modulated. If the amplification coefficient of the loop is large enough, the phase variations are suppressed; thus, only the signal amplitude is modulated.

3.3.5 Tunable Semiconductor Lasers

Semiconductor lasers that can be electronically tuned over a wide range of optical frequencies (wavelengths) play a crucial role in a variety of FDM and WDM multichannel lightwave communications systems. According to Chapter 2, we will assume that FDM systems are coherent lightwave systems, while WDM systems present IM/DD systems. In this section, we are paying attention only to the tuning capabilities of semiconductor lasers. The implications of the laser characteristics on system design will be discussed in Chapter 5 and Chapter 9.

According to the characteristics of FDM and WDM systems, a several terahertz frequency tuning range is required of the tunable lasers [27]. This frequency range corresponds to an 8-nm wavelength range at the 1550-nm wavelength region. The lasers can be tuned either continuously [24] or in discrete steps [28]. While continuously tuned lasers do not have a predefined set of operating wavelengths, discretely tuned lasers have a predefined set of easily accessible wavelengths (frequencies) that define separate optical channels. Besides, the wide tuning range of tunable semiconductor lasers should operate in a single mode with good side-mode suppression. The efficient tunable laser should have access to several tens of frequency channels followed by simple channel access control. These criteria make discretely tuned lasers preferable to continuously tuned lasers for application in WDM systems. While traditional DFB and distributed Bragg reflector (DBR) lasers give a tuning range limited to about 1 THz, some realizations of discretely tuned lasers, such as the so-called *three branch Y3 laser* [29], offer a tuning range of nearly 7 THz and potential access to more than 100 optical channels.

Tunable semiconductor lasers use an intracavity filter that selects an appropriate single lasing mode from the set of possible modes in the resonant cavity. The filter is tuned by an electrically induced change of the refractive index. The relative shift of the optical frequency of the filter transmission peak can be calculated as

$$\left(\frac{\Delta f}{f}\right) = -k\left(\frac{\Delta n}{n}\right) \tag{3.95}$$

where Δf is the frequency shift, $\Delta n/n$ is the relative index change, and k is the so-called lever coefficient [30], which is close to unity for the simple Bragg reflector. The achievable index changes in DFB and DBR lasers are typically less than 1%, and that fact limits the frequency-tuning range in these lasers to a maximum 2 THz in the 1550-nm wavelength region.

To spread the frequency-tuning range, the coefficient k should be as high as possible. This can be achieved in discretely tunable structures, such as that described in [29], where the laser tunes with the filter by jumping from one lasing mode to another. Such a laser structure includes an additional grating structure in the resonant cavity that increases the lever coefficient (see Section 9.3.3 about optical filters).

The most important characteristics of tunable semiconductor lasers are side-mode suppression, intercavity optical filter finesse, and the number of accessible channels. The intercavity filter plays an essential role in wavelength selectivity, and it is supposed to select a single lasing mode from the Fabry-Perot set of possible modes and to produce a side-mode suppression ratio, R, of more than 20 dB. The ratio R can be calculated as [30]

$$R = 1 + \frac{\ell n(T_{sm})}{\ell n(r_1 r_2)} \frac{P_o}{P_{sp}}$$ (3.96)

where r_1 and r_2 are the power reflectivities of the two laser mirrors, T_{sm} is the ratio of the filter transmission between the side mode and the main mode, P_o is the output power, and P_{sp} is the intracavity spontaneous emission power into lasing mode.

If we have periodic intracavity filter, the effective tuning range of the laser is given by the filter period or free spectral range. In discretely tuned lasers, frequency channels are defined by corresponding lasing modes. The number of channels, M, that can be accessible in the tuning range is determined by the channel spacing, δf. The channel spacing, δf, depends on parameter R, that is, a lower R means closer channel spacing. If filter finesse is defined by parameter $F = \Delta f / \Delta f_c$, where Δf defines the filter period and Δf_c is the filter frequency bandwidth, the channel spacing can be expressed as

$$\delta f = \left(\frac{2\Delta f_c}{\pi} \right) \cos^{-1}(\sqrt{T_{sm}})$$ (3.97)

The number of accessible channels is

$$M = \frac{\Delta f}{\delta f} = \frac{\pi \Delta f}{2 \cos^{-1}(\sqrt{T_{sm}})}$$ (3.98)

The number of accessible channels depends strongly on the filter finesse, F. So for $F = 2$, the number of channels is about 20, for $F = 6$, the number of channels is about 80; while for filter finesse near 14, the number of channels is about 100. The number of accessible channels depends on the parameter T_{sm}, as well. Thus, for $T_{sm} = 0.95$ and $T_{sm} = 0.97$ and a filter finesse of about 10, we have a number of channels that varies from 70 to 90, respectively. At the same time, the side-mode suppression ratio varies from a range of 25 to 30 dB down to a range of 22 to 27 dB, respectively, with the output optical power $P_o = 10$ mW. If the output optical power goes down to 1 mW, the side-mode suppression ratio decreases for about 10 dB [27]. As we can conclude, there can be a tradeoff between the required laser side-mode suppression and the number of accessible channels. A larger side mode suppression requires a smaller number of accessible channels and vice versa.

REFERENCES

[1] Born, M., and E. Wolf, *Principles of Optics*, New York: Pergamon Press, 1964.

[2] Morgan, J., *Introduction to Geometrical and Physical Optics*, New York: McGraw-Hill, 1953.

[3] Howes, M. J., and D. V. Morgan, eds., *Optical Fiber Communications, Devices, Circuits, Systems*, New York: John Wiley and Sons, 1982.

[4] Henry, C. H., "Theory of Linewidth of Semiconductor Lasers," *IEEE J. Quantum Electron.*, QE-18(1982), pp. 259–264.

[5] Yamamoto, Y., "AM and FM Quantum Noise in Semiconductor Lasers," *IEEE J. Quantum Electron.*, QE-19(1983), pp. 34–36.

[6] Vahala, K., and A. Yariv, "Semiclassical Theory of Noise in Semiconductor Lasers, Parts I and II," *IEEE J. Quantum Electron.*, QE-19(1983), pp. 1096–1101 and 1102–1109.

[7] Spano, P., et al., "Phase Noise in Semiconductor Lasers: A Theoretical Approach ," *IEEE J. Quantum Electron.*, QE-19(1983), pp. 1195–1199.

[8] Sargent, M., et al., *Laser Physics*, London: Addison-Wesley, 1974.

[9] Henning, I. D., and J. V. Collines, "Measurement of Linewidth Broadening Factor α of Semiconductor Lasers," *Electron. Lett.*, 19(1983), pp. 927–929.

[10] Kikuchi, K., and T. Okoshi, "Estimation of Linewidth Enhancement Factor of GaAlAs Laser by Correlation Measurement Between AM and FM Noises," *IEEE J. Quantum Electron.*, QE-21(1985), pp. 669–673.

[11] Brosson, P., "Analytical Model of Semiconductor Optical Amplifier," *IEEE/OSA J. Lightwave Techn.*, LT-12(1994), pp. 49–54.

[12] Kikuchi, K., and T. Okoshi, "Measurement of FM Noise, AM Noise and Field Spectra of 1.3 μm InGaAsP Lasers and Determination of the Linewidth Enhancement Factor," *IEEE J. Quantum Electron.*, QE-21(1985), pp. 1814–1818.

[13] Merzbacher, E., *Quantum Mechanics*, New York: John Wiley and Sons, 1961.

[14] Papoulis, A., *Probability, Random Variables and Stohastic Processes*, Tokyo: McGraw-Hill, 1965.

[15] Haug, H., "Quantum Mechanical Rate Equation for Semiconductor Lasers," *Phys. Rev.*, 10(1969), pp. 338–348.

[16] Kikuchi, K., and T. Okoshi, "FM and AM Noise Spectra of 1.3 μm InGaAsP DFB Lasers in 0-3 GHz Range and Determination of Their Linewidth Enhancement Factor α," *Electron. Lett.*, 20(1984), pp. 1044–1045.

[17] O'Mahoni, M. J., and I. D. Henning, "Semiconductor Laser Linewidth Broadening Due to 1/f Carrier Noise," *Electron. Lett.*, 19(1983), pp. 1000–1001.

[18] Hunziger, G., et al., "Gain, Refractive Index, Linewidth Enhancement Factor From Spontaneous Emission of Strained GaInP Quantum Well Lasers," *IEEE J. Quantum Electron.*, QE-31(1995), pp. 643–646.

[19] Kikuchi, K., and T. Okoshi, "Simple Formula Giving Spectrum Narrowing Ratio of Semiconductor Laser Output Obtained by Optical Feedback," *Electron. Lett.*, 18(1982), pp. 10–12.

[20] Tanagawa, T., et al., "Frequency Stabilization of 1.5 μm InGaAsP Distributed Feedback Laser on NH_3 Absorption Line," *Appl. Phys. Lett.*, 45(1984), pp. 826–828.

[21] Tench, R. E., et al., "Performance Evaluation of Waveguide Phase Modulators for Coherent Systems at 1.3 and 1.5 μm," *IEEE/OSA J. Lightwave Techn.*, LT-51(1987), pp. 492–501.

[22] Amann, M. C., "Tunable Lasers and Their Applications," *OFC/100C '93*, paper FG, San Jose, 1993.

[23] Yochikuni, Y., and G. Motosugi, "Independent Modulation in Amplitude and Frequency Regimes by Multielectrode Distributed Feedback Laser," *Proc. 9th Conf. Optical Fiber Communications*, Atlanta, 1986, pp. 223–225.

[24] Coch, T. L., and U. Koren, "Semiconductor Lasers for Coherent Communications," *IEEE/OSA J. Lightwave Techn.*, LT-8(1990), pp. 274–293.

[25] Nilson, S., et al., "Improved Spectral Characteristics of MQW-DFB Lasers by Incorporation of Multiple Phase Shift," *IEEE/OSA J. Lightwave Techn.*, LT-13(1995), pp. 434–441.

[26] Seltzer, C. P., et al., "The gain-lever in InGaAsP/InP Multiple Quantum Well Lasers, *IEEE/OSA J. Lightwave Techn.*, LT-13(1995), pp. 283–289.

[27] Tohmori, Y., et al., "Broad-Range Wavelength Tunable Superstructure Grating (SSG) DBR Lasers," *IEEE J. Quantum Electron.*, QE-29(1993), pp. 1817–1823.

[28] Kuznetsov, M., et al., "Widely Tunable (45 nm, 5.6 THz) Multiquantum Well Three Branch Y3 Lasers for WDM Networks," *IEEE Photon. Techn., Lett.*, 5(1993), pp. 879–882.

[29] Kuznetsov, M., et al., "Design of Widely Tunable Semiconductor Three Branch Lasers," *IEEE/OSA J. Lightwave Techn.*, LT-12(1994), pp. 2100–2106.

[30] Kuznetsov, M., et al., "Asymmetric Y-Branch Tunable Semiconductor Laser With 1 GHz Tuning Range," *IEEE Photon. Techn. Lett.*, 4(1992), pp. 1093–1095.

Chapter 4

Optical Receivers for Coherent Lightwave Systems

4.1 INTRODUCTION

The basic parameters related to optical receivers for coherent lightwave transmission systems were introduced in Chapter 1 and Chapter 2. It was pointed out that coherent optical receivers differ from direct detection receivers mainly because detection of the incoming optical signal is performed together with a local optical oscillator signal. Such detection of combined optical signals, consisting of the incoming optical signal and the local oscillator signal, provides suppression of thermal noise influence in the receiver, leading to increased receiver sensitivity. The sensitivity of a coherent optical receiver depends on the applied modulation method in the optical transmitter, the applied detection scheme, and adjustment of the local optical oscillator parameters with the parameters of incoming optical signals.

It was pointed out in Chapter 2 that an FSK modulation scheme provides improved receiver sensitivity compared with an ASK modulation scheme. Further improvement can be achieved if a PSK modulation scheme is applied. Still further improvement in receiver sensitivity can be obtained by the employment of a homodyne detection method instead of a heterodyne one. But there are some difficulties concerning the practical realization of theoretical anticipation, caused by imperfections of the key elements for the chosen detection scheme. That is especially valid for homodyne detection schemes. Further technological improvements will likely eliminate these imperfections.

This chapter will examine the characteristics of the basic elements in a coherent optical receiver. Since the term *coherent optical receiver* is rather broad, some of the variants will be considered in more detail. At the very end of this chapter, some advanced realizations of the coherent optical receiver will be described.

4.2 BASIC STRUCTURE OF COHERENT OPTICAL RECEIVERS

The basic elements of a coherent optical receiver are presented in Figure 4.1. This schematic representation illustrates the general character of a coherent optical receiver. The main characteristics of the coherent receiver are determined by the applied type of demodulator.

The optical signal carrying information comes from the optical fiber line and combines with the signal from the local optical oscillator in the optical mixer. As the optical mixer, either an optical beam splitter (or half-mirror) or an optical directional optical coupler can be used. The main role of the optical mixer is to perform the adjustment between the parameters of incoming optical signals and the parameters of the local oscillator signal; that is, the adjustment of signal directions before they fall on the photodetector. The optical mixer is very sensitive to any mechanical instability, a fact that must be taken into account in optical receiver design.

The relations between the frequencies and the phases of incoming optical signals and local oscillator signals depend on the type of applied detection scheme. In a heterodyne receiver, a difference-frequency signal, called an intermediate frequency (IF) signal, is obtained after the mixing and photodetection processes. In a homodyne optical receiver, the phase and the frequency of the local oscillator are controlled to be equal to the frequency and the phase of the incoming optical signal. Thus, the IF in a homodyne optical receiver is equal to zero. That is why in a heterodyne optical receiver there is no need for any phase matching. The central frequency of the IF filter must be equal to the frequency shift between the incoming optical signal and the local optical oscillator signal. The phase and the frequency adjustments in a homodyne optical receiver are complex operations. A critical technique, such as a PLL or phase diversity detection, should be applied to achieve the matching of frequency and phase.

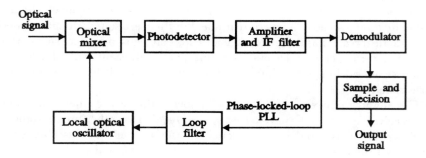

Figure 4.1 The basic structure of a coherent optical receiver.

The combined optical signal from the optical mixer falls on the photodetector. In the specific design, the so-called *balanced coherent receiver*, the optical signal from the mixer is divided into two parts, and each is photodetected by a following photodiode. Thus, two independent photodetectors are present, generating two information electrical signals, but the signals' difference is relevant for the subsequent steps in the receiver.

The photodetector performs not only an optoelectronic conversion but also the efficient driving out of the frequencies in the combined optical signal, due to its nonlinear characteristics. The PIN photodiodes and the APDs are, in fact, the only types of photodetectors used in a coherent receiver. There are many reasons for that such as high efficiency, small geometry, and fast response. As an illustration, we can say that quantum efficiency of quality semiconductor photodiodes are above 80% at the operating wavelength. The silicium-based photodiodes are the most efficient in the spectral region of 0.8 to 0.9 μm, while InGaAsP photodiodes possess the most favorable characteristics in the spectral regions corresponding to the second and third spectral windows, around wavelengths 1.3 μm and 1.55 μm, respectively.

The impulse response of the photodiode employed in a coherent optical receiver must be, in general, very narrow; consequently, the corresponding frequency bandwidth should be as wide as possible. The photodiode employed in a heterodyne optical receiver should, in general, have wider frequency bandwidth than that employed in a homodyne optical receiver. So the upper frequency of the photodiode response should not be lower than the modulation signal bandwidth in a homodyne optical receiver, while the heterodyne receiver requires the upper frequency of photodiode response to exceed the fifth-time value of the modulation signal bandwidth [1]. That is the essential advantage of a homodyne detection scheme over a heterodyne detection one.

The receiver amplifier following the photodiode plays a key role in the optical receiver in terms of SNR and receiver sensitivity. The basic requirements for the coherent receiver amplifier are practically the same as those for an amplifier in a DD optical receiver. Thus, we will follow the basic principles of the receiver amplifier design, with necessary corrections due to local oscillator influence [2–5].

The demodulator following the receiver amplifier in a coherent optical receiver has, in general, the same characteristics as the demodulator in a classical microwave system. Since the detection schemes applied for different kinds of modulation were already considered in Chapter 2, the main attention in this chapter will be to the coherent procedures used in the demodulator and to the application of an electrical PLL. The design of the demodulator determines the type and the complexity of the coherent optical receiver (ASK, FSK, PSK, or DPSK). For example, an ASK demodulator has the simplest structure (see Figure 2.3). If necessary, there may be an equalizer before the demodulator in the coherent receiver scheme, playing the same role as the one in a DD optical receiver. Finally, the FSK and PSK demodulators must include the corresponding discriminators for FSK-ASK and PSK-ASK conversions, respectively, before the decision process.

4.2.1 Characteristics of Coherent Receiver Amplifiers

It is well known that the front-end amplifier has the decisive role in the thermal-noise generation process in a DD optical receiver. With the use of a local optical generator, the influence of thermal noise is suppressed, but the front-end amplifier in a coherent optical receiver should be carefully designed. We will consider the characteristics of a front-end amplifier from the point of view of thermal-noise suppression and the minimization of total noise.

The equivalent circuit of the photodiode and the front-end amplifier is given in Figure 4.2. The equalizer following the amplifier is not included in this scheme. The photodiode from Figure 4.2 is represented by the current generators of signal and quantum noise, connected parallel with photodiode capacity C_d and external load resistance R_L. The voltage produced by the current generators on the resistance R_L is amplified by the front-end amplifier.

The front-end amplifier is represented by an input impedance (consisting of input resistance and input capacity), by current and voltage generators of thermal noise, and by the stage with an ideal amplification, A. In accordance with (2.16), the average value of the current signal from the current-signal generator in Figure 4.2 can be expressed as

$$\overline{i} = 2\frac{\eta q}{hf}M\sqrt{P_sP_L} = 2\Re M\sqrt{P_sP_L} \qquad (4.1)$$

where η is the quantum efficiency of photodiode, q is the electron charge, h is Planck's constant, M is the avalanche gain coefficient, P_s is an average value of incoming

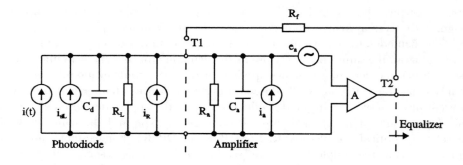

Figure 4.2 The equivalent circuit of the photodiode and front-end amplifier.

optical power, P_L is the average value of local optical oscillator power, and $\mathfrak{R}$ is the photodiode responsivity. Equation (4.1) is valid both for a PIN photodiode ($M = 1$) and for an APD ($M > 1$).

The current-noise generator from Figure 4.2 gives two components, referring to the shot noise of an incoming signal and the shot noise of a local oscillator signal. In accordance with (2.11) and (2.16) and Appendix F, the power of total shot noise at the output of a heterodyne receiver amplifier is

$$\overline{i_{sL}^2} = \overline{i_s^2} + \overline{i_L^2} = 2q\left(\frac{\eta q}{hf}\right)M^2F(M)(P_s + P_L)2B \tag{4.2}$$

where $F(M)$ is the excess noise factor of the photodiode (see Appendix F). The equivalent frequency bandwidth of the amplifier is limited to $2B$ for heterodyne detection and to B for homodyne detection, where B is the modulating signal bandwidth. The excess noise factor function, $F(M)$, may have the forms given in Appendix F, so we chose the most appropriate model, $F(M) = M^x$, because our considerations have the general character.

The shot noise current generates the noise voltage at the output of the amplifier, having the power given as

$$\overline{\mathcal{V}_{sL}^2} = \overline{i_{sL}^2}R^2A^2 \tag{4.3}$$

where

$$\frac{1}{R} = \frac{1}{R_a} + \frac{1}{R_L} \tag{4.4}$$

The value $1/R$ is the equivalent conductivity of the scheme from Figure 4.2. At the same time, the equivalent capacity is

$$C = C_d + C_a \tag{4.5}$$

Due to the existence of the local optical oscillator, the influence of dark current noise (considered in Chapter 2) can be neglected, so the total noise power at the output of the front-end amplifier can be expressed as

$$\mathcal{V}_n(t) = [\mathcal{V}_{sL}^2(t) + \mathcal{V}_R^2(t) + \mathcal{V}_I^2(t) + \mathcal{V}_E^2(t)]^{1/2} \tag{4.6}$$

where $\mathcal{V}_R(t)$ is the noise voltage on the load resistance, R_L, having the nature of the thermal noise, while $\mathcal{V}_I(t)$ and $\mathcal{V}_E(t)$ are the noise voltages generated by the current and the voltage noise generators, respectively.

It can be assumed that noise sources are mutually independent [2], so the total output noise power can be presented as a sum of statistically independent noises, or

$$\overline{\mathcal{V}_n^2(t)} = \overline{\mathcal{V}_{sL}^2(t)} + \overline{\mathcal{V}_R^2(t)} + \overline{\mathcal{V}_I^2(t)} + \overline{\mathcal{V}_E^2(t)} \tag{4.7}$$

To evaluate the total noise power, the terms on the right side of (4.7) should be evaluated first.

The first term on the right side of (4.7) has the form

$$\overline{\mathcal{V}_{sL}^2(t)} = 2q\left(\frac{\eta q}{hf}\right)M^{x+2}(P_s + P_L)(2B)R^2A^2 \tag{4.8}$$

The thermal noise power on the resistance, R_L, is given by the known relation

$$\overline{i_R^2(t)} = \frac{4k\Theta}{R_L} \tag{4.9}$$

where k is Boltzmann's constant and Θ is the absolute temperature. The mean square value of the amplifier output voltage due to thermal noise can be presented as

$$\overline{\mathcal{V}_R^2(t)} = \frac{4k\Theta}{R_L}(2B)R^2A^2 \tag{4.10}$$

Since we assumed that current and voltage sources of the noise are independent in the front-end amplifier, they can be represented by their own spectral densities, denoted by G_I and G_E, respectively. The mean square value of the noise voltage, generated by the current source, at the amplifier output is given as

$$\overline{\mathcal{V}_I^2(t)} = 2G_I(2B)R^2A^2 \tag{4.11}$$

Since the voltage source of noise is in a serial connection, (4.11) cannot be directly applied to it. Hence, the voltage source of noise should be considered as another parallel branch. The corresponding transformation of the spectral density, G_E, must be made after insertion of the voltage source in another parallel branch. The corresponding current source of the noise in a parallel branch will have the noise spectral density equal to

$$G_E' = \int_0^\infty |Y(f)|^2 G_E df \tag{4.12}$$

where $Y(f)$ is the total admittance of the circuit from Figure 4.2, given as

$$Y(f) = 1/R + j2\pi fC \tag{4.13}$$

By the integration from 0 to 2π in (4.12), it is obtained that

$$G'_E = G_E\left[\frac{2B}{R^2} + (2\pi C)^2\frac{(2B)^3}{3}\right] \tag{4.14}$$

The mean square value of the noise voltage, generated by a voltage source of noise at the output of the amplifier, is now given as

$$\overline{\mathcal{V}_E^2(t)} = 2G'_E R^2 A^2 \tag{4.15}$$

Since the values of all the terms on the right side of (4.7) are known, we can write the final expression for the mean square value of the noise voltage generated at the output of the front-end amplifier. Before that, we can use the following unified denotation for frequency bandwidth:

$$B' = \begin{cases} B \text{ for homodyne detection} \\ 2B \text{ for heterodyne detection} \end{cases} \tag{4.16}$$

The mean square value of the noise voltage at the output of coherent receiver front-end amplifier is

$$\overline{\mathcal{V}_n^2(t)} = \left\{2q\left[\frac{\eta q}{hf}\right]M^{x+2}[P_s + P_L] + \frac{4k\Theta}{R_L} + 2G_I\right\}B'R^2A^2 \tag{4.17}$$
$$+ \left[\frac{B'}{R^2} + (2\pi C)^2\frac{(B')^3}{3}\right]2G_E R^2 A^2$$

Equation (4.17) can be expressed as

$$\overline{\mathcal{V}_n^2(t)} = \overline{\mathcal{V}_{nq}^2(t)} + \overline{\mathcal{V}_{nt}^2(t)} \tag{4.18}$$

where

$$\overline{\mathcal{V}_{nq}^2(t)} = \left[2q\left(\frac{\eta q}{hf}\right)M^{x+2}(P_s + P_L)\right]B'R^2A^2 \tag{4.19}$$

and

$$\overline{\mathcal{V}^2_{nt}(t)} = \left(\frac{4k\Theta}{R_L} + 2G_I\right)B'R^2A^2 + \left[\frac{B'}{R^2} + (2\pi C)^2\frac{(B')^3}{3}\right]2G_E R^2 A^2 \qquad (4.20)$$

are the components denoting contributions of the quantum and thermal noises, respectively.

Since the incoming optical signal, according to (4.1), generates the voltage $\mathcal{V} = \overline{i}RA$ at the output of the front-end amplifier, the SNR at the output of amplifier can be expressed as

$$SNR = \frac{4\mathcal{R}^2 A^2 R^2 P_s P_L}{\overline{\mathcal{V}^2_{nq}(t)} + \overline{\mathcal{V}^2_{nt}(t)}} \qquad (4.21)$$

This ratio is the concrete form of (2.17) and (2.18), so it can be used to evaluate the coherent optical receiver sensitivity for different modulation schemes.

The contributions of the quantum and the thermal noises can be considered for the common realizations of front-end amplifiers, which we will do next to evaluate the SNR and the total transmission capacity.

4.2.2 Front-End Amplifier Realizations

4.2.2.1 FET-Based Amplifiers

The power spectral density of the input current noise source in the field effect transistor (FET)–based amplifier is [5]

$$G_{I,FET} = \frac{4k\Theta}{R_a} \qquad (4.22)$$

It is well known that FET has a high input resistance, R_a, usually higher than 1 MΩ, so it can be assumed that $R_a \to \infty$. In such a case, the total resistance from Figure 4.2 is nearly equal to R_L, and the power spectral density of the current-source noise can be neglected.

The thermal noise due to resistance of the conducting channel dominates in FETs. This noise is characterized by the transconductance g ($g \simeq 5 \cdot 10^{-3}\Omega^{-1}$) [3, 6]. Such a noise belongs, in fact, to the voltage source of noise from Figure 4.2, and has the power spectral density

$$G_{E,FET} = \frac{2}{3}\frac{4k\Theta}{g} \qquad (4.23)$$

The thermal component of the total noise at the output of a front-end amplifier with FET is obtained from (4.20) and has the form

$$\overline{\mathcal{V}_{nt}(t)} = \frac{4k\Theta}{R_L}B'R^2A^2 + \left[\frac{B'}{R_L^2} + (2\pi C)^2\frac{(B')^3}{3}\right]\frac{4}{3}\frac{4k\Theta}{g}R^2A^2 \qquad (4.24)$$

To minimize the thermal noise value, the resistance R_L must be as large as possible. But such a choice will lead to an integration effect of total current signal due to the resistance-capacity (RC) factor, in spite of the fact that total capacity is rather small (usually below 5 pF). The integration effect can be eliminated by insertion of a differentiator after the front-end amplifier.

Only GaAs FET can be used practically in the amplifier realization of a coherent optical receiver, because these transistors have very fast responses and provide the possibility for an integrated optoelectronic realization of several elements (for example, PIN-FET integrated realization).

4.2.2.2 Bipolar Transistor Amplifiers

The power spectral density of the thermal-noise current generator in a bipolar transistor-based front-end amplifier is given as [6]

$$G_I = 2qI_b = \frac{2k\Theta}{R_{in}} \qquad (4.25)$$

where I_b is the base current of the transistor, and R_{in} is its input resistance. The resistors in the base circuit of the transistors are chosen to have very large values, so the following relation is valid:

$$R_a = R_{in} = \frac{k\Theta}{qI_b} \qquad (4.26)$$

This, in fact, means that the thermal component of the total noise is determined by the choice of the transistor.

The power spectral density of thermal noise generated by the current source from Figure 4.2 in the bipolar transistor front-end design is given as

$$G_E = \frac{2k\Theta}{g} \qquad (4.27)$$

where g is the conductance determined as the ratio of the current amplification, β, of the transistor and its input resistance, or

$$g = \frac{\beta}{R_{in}} = \frac{I_c}{I_b R_{in}} = \frac{qI_c}{k\Theta} \tag{4.28}$$

(I_c is the collector current.)

By the replacement of (4.25), (4.26), (4.27), and (4.28) into (4.20), the thermal component of total noise at the output of the front-end amplifier realized by bipolar transistors in a coherent optical receiver becomes

$$\overline{V_{nt}^2(t)} = \left[\frac{4k\Theta}{R_L} + \frac{4k\Theta}{R_{in}}\right]R^2A^2 + \left[\frac{B'}{R^2} + (2\pi C)^2\frac{(B')^3}{3}\right]\frac{4k\Theta}{g}R^2A^2 \tag{4.29}$$

The value in (4.29) can be decreased by an increase of resistance R_L, but an integration effect will appear again.

4.2.2.3 Transimpedance Front-End Amplifiers

If an electrical feedback with resistance R_f, as in Figure 4.2, is made, a so-called *transimpedance amplifier* is obtained. The transimpedance amplifier offers larger frequency bandwidth than the realizations with high input resistance. The transimpedance amplifier has high input resistance also, but its frequency bandwidth is wider than that of classical high-resistance realizations.

The characteristics of the transimpedance amplifier can be easily analyzed by (4.20), if the load resistance, R_L, is replaced by the feedback resistance value

$$R_{Lf} = \frac{R_L + R_f}{R_L R_f} \tag{4.30}$$

In such a way, (4.20) becomes

$$\overline{V_{nt}^2(t)} = \left(\frac{4k\Theta}{R_{Lf}} + 2G_I\right)B'R^2A^2 + \left[\frac{B'}{R_1^2} + (2\pi C)^2\frac{(B')^3}{3}\right]2G_E R^2A^2 \tag{4.31}$$

where

$$\frac{1}{R_1} = \frac{1}{R} + \frac{1}{R_f} = \frac{1}{R_a} + \frac{1}{R_L} + \frac{1}{R_f} \tag{4.32}$$

Since in practical situations the feedback resistance is much higher than the input resistance, we have that $R_1 \simeq R$, and the mean square value of the noise voltage at the amplifier output can be expressed as the sum

$$\overline{\mathcal{V}_{nt}^2(t)} = \overline{\mathcal{V}_{nt,H}^2(t)} + \frac{4k\Theta}{R_f}B'R^2A^2 \tag{4.33}$$

where the first term on the right side belongs to the amplifier with high input impedance, while the second term presents the "contribution" due to feedback. The feedback contribution decreases the total sensitivity of the optical receiver, which is the only imperfection of such a realization, compared with the realizations with high input impedance. However, this sensitivity decrease is not so large and can be accepted.

The transfer function of the transimpedance amplifier, in accordance with Figure 4.2, has the form

$$H_T = \frac{1}{1 + j2\pi fRC/A} \tag{4.34}$$

with corresponding bandwidth equal to $A/(4RC)$. These results can be compared with ones corresponding to the high-impedance amplifier. Since the transfer function of the amplifier without feedback is equal to

$$H_H = \frac{AR}{1 + j2\pi fRC} \tag{4.35}$$

it follows that its bandwidth is equal to $1/(4RC)$. Hence, the transimpedance amplifier bandwidth is considerably higher than the bandwidth of high-impedance amplifier realizations.

4.2.3 Demodulator Characteristics in Coherent Optical Receivers

4.2.3.1 Demodulation Methods for ASK, FSK, and PSK Signals

Demodulation methods employed for ASK, FSK, and PSK signals were considered in Chapter 2. The demodulation procedure depends on the following facts (among others): whether only one or two photodiodes are used in the photodetection process; whether detection has a coherent character or an envelope detection is made; and whether one or two IF filters are used (in FSK receiver).

In this section, we will assume that only one photodiode is used and that photodetection has the coherent character. Hence, there is a local electrical oscillator with the oscillation frequency equal to the frequency shift between the incoming optical signal and the local optical oscillator signal. It is clear that such a scheme is valid only for heterodyne detection. In fact, the heterodyne detection scheme includes two local oscillators: an optical local oscillator and an electrical local oscillator.

Homodyne detection is imposed only by an optical local oscillator, because the frequency shift between the incoming optical signal and the local optical oscillator signal does not exist.

Since the electrical PLL is used in the heterodyne optical receiver, we will consider its influence on the detected signal characteristics and discuss the phase shift behavior under the PLL influence. After that, the characteristics of the dual-photodetector coherent optical receivers, or phase diversity receivers, will be analyzed.

4.2.3.2 Influence of Electrical PLL on Total Noise

To demodulate an amplified signal passed through the IF band filter in Figure 4.1, it is necessary to have a referent electrical signal with frequency equal to the central frequency of the IF filter. The referent electrical signal is generated by an electrical PLL and must follow only the carrier component of the signal at the output of the IF filter. The electrical PLL operation can be explained on the simplified scheme shown in Figure 4.3.

The principal elements of the electrical PLL are the phase detector and the voltage-controlled oscillator (VCO), which is in the feedback branch. The carrier wave with frequency equal to the frequency shift between the incoming optical signal and the local optical oscillator signal leaves the IF filter and enters the multiplicator. Since both the frequency and the phase of the IF signal can fluctuate around some mean values, the IF signal can be written in the form

$$s_1(t) = S_1 \sin \psi_1(t) = S_1 \sin[\omega_1(t)t + \varphi_1(t)] \qquad (4.36)$$

while the referent wave can be written as

$$s_r(t) = S_2 \cos \psi_2(t) = S_2 \sin[\omega_2(t)t + \varphi_2(t)] \qquad (4.37)$$

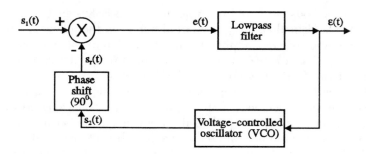

Figure 4.3 The schematic representation of electrical PLL.

If the feedback is broken, or if the operating voltage at the VCO input is equal to zero, the output signal from the VCO is equal to

$$s_2(t) = S_2 \sin(\omega_2 t + \varphi_2) \tag{4.38}$$

The frequency, ω_2, presents a stationary value of the VCO. If the feedback in the PLL is connected, the oscillation frequency will, in general, differ from the value of ω_2. That difference is proportional to the operating voltage, $\epsilon(t)$, which in such a case presents the filtrated product of the incoming signal and the referent wave. Only the low frequencies around the difference $\omega_1 - \omega_2$ are allowed to pass through the filter, so we have

$$\epsilon(t) = K_m s_1(t) s_2(t) \left.\begin{matrix} {\scriptstyle \omega = \omega_1 - \omega_2} \\ \\ {\scriptstyle \omega = 0} \end{matrix}\right. \tag{4.39}$$

where K_m is a multiplication constant, and the vertical line symbolizes the filtration process.

The instantaneous frequency of the voltage from the VCO is the first derivative of the instantaneous phase, that is,

$$\frac{d\psi_2(t)}{dt} = \omega_2 + K_2 \epsilon(t) \tag{4.40}$$

where K_2 is a dimensional constant. To find the value of the operating voltage, or the phase error, $\epsilon(t)$, it is necessary to find the signal, $e(t)$, at the multiplicator output:

$$e(t) = \frac{1}{2} K_m S_1 S_2 \sin[\psi_1(t) - \psi_2(t)] = K_d \sin[\psi_1(t) - \psi_2(t)] \tag{4.41}$$

If the impulse response of the filter is $h(t)$, the error signal at the filter output can be expressed as

$$\epsilon(t) = \int_0^t e(u) h(t - u) du, \quad t \geq 0 \tag{4.42}$$

By inserting the value of (4.42) into (4.40), we obtain an intergo-differential equation, that is,

$$\frac{d\psi_2(t)}{dt} = \omega_2 + K_2 K_d \int_0^t h(t - u) \sin[\psi_1(u) - \psi_2(u)] du \tag{4.43}$$

The constant ω_2 can be eliminated from (4.37), (4.38), (4.39), and (4.40) by the substitutions

$$\varphi_1(t) = \psi_1(t) - \omega_2 \tag{4.44}$$

$$\varphi_2(t) = \psi_2(t) - \omega_2 \tag{4.45}$$

and (4.43) takes the form

$$\frac{d\varphi_2(t)}{dt} = K_2 K_d \int_0^t h(t - u)\sin[\varphi_1(u) - \varphi_2(u)]du \tag{4.46}$$

Finally, if the phase difference is denoted by

$$\varphi(t) = \varphi_1(t) - \varphi_2(t) \tag{4.47}$$

the final form of the characteristic equation of the electrical PLL is obtained:

$$\frac{d\varphi(t)}{dt} = \frac{d\varphi_1(t)}{dt} - K_2 K_d \int_0^t h(t - u)\sin\varphi(u)du \tag{4.48}$$

Equation (4.48) can be represented by an equivalent scheme of PLL, shown in Figure 4.4. The multiplicator is now replaced by a collector and sine nonlinearity, while the VCO is replaced by an integrator, which is excited by the derivative of the phase $\varphi_2(t)$. The total amplification in the loop is now equal to half the product of the signal amplitude, the multiplication constant, and the VCO constant.

The equivalent scheme from Figure 4.4 can be simplified additionally, if the phase difference, $\varphi(t)$, is very small during the considered time interval (usually

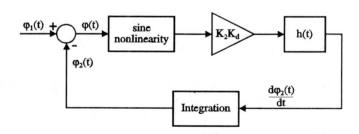

Figure 4.4 Equivalent model of electrical PLL.

lower than $\pi/6$). Then the approximation $\sin \varphi \simeq \varphi$ can be applied, and the sine nonlinearity can be neglected. Such a case corresponds to the linear model of PLL, which is more suitable for further analysis.

In such a case, (4.48) takes the form

$$\frac{d\varphi(t)}{dt} + K_2 K_d \int_0^t h(t - u)\varphi(u)du = \frac{d\varphi_1(t)}{dt} \tag{4.49}$$

By applying the Laplace transform to (4.49), the following equation is obtained:

$$s\Phi(s) + K_2 K_d H(s)\Phi(s) = s\Phi_1(s) \tag{4.50}$$

where $\Phi(s)$, $\Phi_1(s)$, and $H(s)$ are the Laplace transforms of functions $\varphi(t)$, $\varphi_1(t)$, and $h(t)$, respectively. Thus, the Laplace transform of the phase difference function becomes

$$\Phi(s) = \frac{s}{s + K_d K_2 H(s)}\Phi_1(s) \tag{4.51}$$

The Laplace transform of the VCO instantaneous phase is

$$\Phi_2(s) = \frac{K_d K_2 H(s)}{s + K_d K_2 H(s)}\Phi_1(s) \tag{4.52}$$

The transfer function $H_P(s)$ of the PLL, given by the ratio $\Phi_2(s)/\Phi_1(s)$, is

$$H_p(s) = \frac{K_d K_2 H(s)}{s + K_d K_2 H(s)} \tag{4.53}$$

The Laplace transforms of the phase difference and the instantaneous input phase are connected by the equation

$$\Phi(s) = [1 - H_p(s)]\Phi_1(s) \tag{4.54}$$

Hence, the shape of the transfer function of the PLL depends primarily on the shape of the filter transfer function. If the filter transfer function is the ratio of two polynomials, then the PLL transfer function will be a ratio of two polynomials, as well. The PLL transfer function will describe the stable system if its poles lie in the left half-plane of the s-complex plane. The order of the PLL is determined by the polynomial degree in the denominator. In most situations, only the PLL of first, second, and third order can be employed.

PLL behavior can be analyzed by performing the inverse Laplace transform of the transfer function and by observation of the time response behavior in a long time interval. Thus, the PLL of first order, which does not contain the filter, cannot reach the absolute synphase state characterized by zero phase difference. The stationary phase difference in the first-order PLL has the finite value, which is reciprocally proportional to the amplification in the loop. It means that the VCO oscillates synchronously but not synphasely with the incoming carrier wave.

The second-order PLL contains the filter with the characteristics of an integrator; hence, the phase difference between the incoming carrier wave and the local wave from VCO will be zero, and the total synphase state is reached. In practical design, however, there is not an ideal filter, so the phase difference will not be zero, but some definite value. The proper design of the loop filter minimizes such a stationary phase difference.

The third-order PLL, realized by two nonideal integrators, provides the zero value of phase difference, but this PLL can fall into an unstable state if the amplification in PLL decreases below some critical value. Thus, the third-order PLL possesses both an advantage and an imperfection compared with the second-order PLL. Because of that, the properly adjusted second-order PLL is commonly used in practical realizations of digital receivers.

To perform an entire analysis of PLL behavior, the influence of the additive noise on PLL characteristics should be considered. It can be assumed that the IF filter, which precedes the PLL and has the central frequency ω_c, allows passing only of noise components with frequencies below the value $2\omega_c$. We can assume that a random process, $f_1(t)$, occurs at the input of the PLL. That random process consists of the incoming signal $s_1(t)$, according to (4.36), and of filtrated noise:

$$n(t) = x(t)\sin \omega_c t - y(t)\cos \omega_c t \qquad (4.55)$$

It is real to suppose that the stationary value of the VCO frequency is equal to the central frequency of the IF filter, so it can be written that

$$f_1(t) = s_1(t) + n(t) = S_1 \sin[\omega_c t + \varphi_1(t)] + x(t)\cos \omega_c t - y(t)\sin \omega_c t \qquad (4.56)$$

In accordance with (4.37), the referent wave now takes the form

$$s_r(t) = S_2 \cos \psi_2(t) = S_2 \cos[\omega_c t + \varphi_2(t)] \qquad (4.57)$$

where $\varphi_2(t)$ has the random character due to noise influence.

The signal at the output of the multiplicator in Figure 4.4 will be

$$e(t) = \frac{1}{2}K_m S_2\{S_1 \sin[\varphi_1(t) - \varphi_2(t)] + x(t)\cos \varphi_2(t) + y(t)\sin \varphi_2(t)\} \qquad (4.58)$$

At the same time, the signal of phase error at the PLL filter output will be

$$\epsilon(t) = K_d \int_0^t \left\{ \sin[\varphi_1(u) - \varphi_2(u)] + \frac{x(u)\cos \varphi_2(u)}{S_1} + \frac{y(u)\sin \varphi_2(u)}{S_1} \right\} \cdot h(t-u)du$$

(4.59)

The first term under the integral presents the phase difference, according to (4.47), while the second term is the equivalent filtrated noise, $n_e(t)$, [4] given as

$$n_e(t) = \left[\frac{x(t)\cos \varphi_2(t)}{S_1} + \frac{y(t)\sin \varphi_2(t)}{S_1} \right]$$

(4.60)

Then, (4.59) can be transformed to

$$\frac{d\varphi(t)}{dt} = \frac{d\varphi_1(t)}{dt} - K_2 K_d \int_0^t h(t-u)[\sin \varphi (u) + n_e(u)]du$$

(4.61)

When the noise at the input of the PLL is a stationary Gaussian process, the equivalent filtrated noise will be a Gaussian process only if the phase, $\varphi_2(t)$, of the VCO is independent of the noise at the PLL input. Since the input noise will practically always exert influence on the instantaneous phase of the VCO, $n(t)$ and $\varphi_2(t)$ will not be entirely independent. To simplify the analysis, however, it is commonly assumed that these processes are independent, since the loop bandwidth is much smaller than the spectral width of the noise. Under such a condition, it is easy to find the mean value and the variance of the equivalent filtrated noise, which are, respectively,

$$\overline{n_e(t)} = 0$$

(4.62)

and

$$\sigma_e^2 = \frac{\sigma^2}{S_1^2}$$

(4.63)

When the carrier wave with the frequency ω_1 and phase φ_1 is led to the PLL input, the phase difference will tend to zero or a finite constant value, depending on the PLL order. But such a conclusion is for PLL behavior without noise influence. The noise influence will cause phase fluctuations around some average value. These fluctuations are called the *phase jitter*. When the noise influence is large enough, it can happen that the instantaneous phase of the VCO makes the step change for

$k(2\pi)$ radians (k is an integer). Hence, the equivalent noise generates the phase jitter and sometimes causes the phase jump for one or a few cycles.

All that can have serious consequences on the error probability in the transmission system. Very strong noise in the PLL causes loss of control in the VCO, since VCO frequency becomes arbitrarily distant from the phase of the incoming wave. Such a situation causes degradation of the PLL function and, consequently, total degradation of transmission quality. Degradation of the PLL function in a coherent optical receiver can be caused by uncontrolled changes of the optical signal polarization state during its propagation through the optical fiber line. This is a serious problem and must be prevented, either by use of polarization-preserving optical fibers or by application of special polarization controllers. Another, more promising method for prevention of random polarization changes is application of the diversity technique in the coherent optical receiver.

The described model for the analysis of noise influence on PLL behavior also can be applied for the analysis of PLL behavior in the detection of signals with different modulations. Namely, the amplitude, frequency, and phase modulations also could be considered as additive disturbances deteriorating the rise of phase coherence. In the case of amplitude modulation of the carrier wave, there is the following signal at the input of PLL:

$$s_1(t) = [S_1 + m(t)]\sin \psi_1(t) \qquad (4.64)$$

where $m(t)$ is a modulating signal. Under the action of the referent wave, according to (4.37), the signal at the output of the multiplicator has the form

$$e(t) = \frac{1}{2}K_m S_2[S_1 + m(t)]\sin[\varphi_1(t) - \varphi_2(t)] \qquad (4.65)$$

while the corresponding signal of phase error will be

$$\epsilon(t) = K_d \int_0^t \left[\sin \varphi(u) + \frac{m(u)\sin \varphi(u)}{S_1} \right] h(t - u)du \qquad (4.66)$$

The PLL operation, when an amplitude-modulated signal is coming at the PLL input, can be described by

$$\frac{d\varphi}{dt} = \frac{d\varphi_1}{dt} - K_2 K_d \int_0^t \left[\sin \varphi(t) + \frac{m(u)\sin \varphi(u)}{S_1} \right] h(t - u)du \qquad (4.67)$$

By comparing (4.67) with (4.61), relative to the influence of the signal and the additive noise on the PLL, we can see that they will have equal forms if the equivalent filtrated noise is replaced by the equivalent modulating signal, $m_e(t)$, where

$$m_e(t) = \frac{m(t)\sin \varphi(t)}{S_1} \qquad (4.68)$$

Equations (4.67) and (4.68) can be used in different situations for the analysis of PLL behavior in real receivers.

Exact analysis of the simultaneous acting of modulation disturbance and additive noise is not possible because of the sine-shape nonlinearity in the PLL model. Because of that, the linear model of PLL is often used, and the linear supposition principle is applied. That principle means the independent analysis of the modulation disturbance and the additive noise and then the summation of their influences [7]. The linear model of PLL is obtained by substituting $\sin\varphi = \varphi$ in (4.61). In the linear model, the equivalent filtrated noise, $n_e(t)$, is directly added to the instantaneous phase, $\varphi_1(t)$, of the input signal. In accordance with that, (4.52) can be directly applied to the Laplace transform, $N_e(s)$, of equivalent filtrated noise. Thus, the spectral power density of phase, $\Phi_2'(f)$, at the output of the VCO is directly related with the spectral power density of equivalent filtrated noise, $\Phi_e'(f)$, or

$$\Phi_2'(f) = |H_p(j2\pi f)|^2 \Phi_e'(f) \qquad (4.69)$$

The spectral power density of the VCO phase is determined by the noise acting, since the instantaneous phase of the VCO fluctuates around its average value under the noise influence. The statistical measure of these fluctuations is the phase variance, given as

$$\sigma_2^2 = \int_0^\infty \Phi_2'(f) df = \int_0^\infty |H_p(j2\pi f)|^2 \Phi_e'(f) df \qquad (4.70)$$

Equation (4.70) is obtained by application of the linear supposition principle, assuming that only noise contributes to the variance of instantaneous phase of the VCO. The modulation and other disturbances cause their own partial variances, which should be added to the variance σ_2^2. Such a principle will be used later on in the analysis of the influence of the optical source on the phase variance.

Equations (4.69) and (4.70) were obtained under the supposition that only the noise influence is present, or that $\varphi_1(t) = 0$, so it follows that the phase difference is

$$\varphi(t) = \varphi_2(t) \qquad (4.71)$$

At the same time, the variance of the phase difference is equal to the phase variance of the VCO, or

$$\sigma_\varphi^2 = \sigma_2^2 \qquad (4.72)$$

Equation (4.70) is essential for determining the value of phase variance in the coherent optical receiver. It is important to consider the dependence of such a variance on the optical-source phase fluctuations (phase noise) or to consider the transformation of optical-source phase noise into receiver phase fluctuations. The determination of phase-noise variance in the coherent optical receiver is essential in exactly evaluating the BER [see (2.58), (2.87), and (2.119), with the transcriptions $\vartheta_n \to \varphi$ and $\sigma_n \to \sigma_\varphi$].

The phase variance evaluation according to (4.70) is, in general, rather complex. The evaluation is considerably simplified with the assumption that the noise at the input of the PLL has the nature of white noise with the constant value of spectral power density (equal to $2\nu_0$). In such a way, the variance of phase difference is

$$\sigma_\varphi^2 = \frac{4\nu_0}{S_1^2} \int_0^\infty |H_p(j2\pi f)|^2 df = \frac{4\nu_0 B_p}{S_1^2} \tag{4.73}$$

where B_p is the equivalent bandwidth of the noise in the PLL (see Appendix B).

The signal-to-noise powers ratio at the PLL input has the value

$$\rho_1^2 = \frac{S_1^2}{2\sigma_1^2} \tag{4.74}$$

while its value at the PLL output, if the input noise is the white noise with the power $2\nu_0 2B_p$, is equal to

$$\rho_p^2 = \frac{S_1^2}{8\nu_0 B_p} = \frac{1}{2\sigma_\varphi^2} \tag{4.75}$$

Equation (4.75) can be applied in different practical situations.

The influence of noise at the PLL input, which causes the fluctuations of the VCO phase, can be replaced by the phase fluctuations of the input signal. Hence, the PLL operation will be disturbed at any rate, either under the influence of equivalent filtrated noise with variance σ_e^2 or under the influence of the input signal-phase fluctuations with variance σ_1^2. Accordingly, the following equality can be written:

$$\sigma_e^2 = \sigma_1^2 = \sigma^2/S_1^2 \tag{4.76}$$

When the SNR in the PLL is very small, an unstable operation of PLL will arise, and phase jumping in the VCO can appear. In such a case, the phase coherency can be broken, and the phase difference can obtain a considerably larger value. In such a situation, the linear model of PLL is not valid anymore, and a nonlinear model should be applied.

The nonlinear model, such as that in Figure 4.3, can be applied if the phase difference is expressed as

$$\varphi(t) = [\varphi_1(t) - \varphi_2(t)]\mathrm{mod}(2\pi) \tag{4.77}$$

Equation (4.77) means that the random variable $\varphi(t)$ remains in the range $[-\pi$ to $\pi]$ for any shift between the input-signal phase and the VCO phase. The behavior of a nonlinear PLL can be considered only if we know the probability density function of the phase difference (or phase error) in the PLL.Thus, an appropriate model for this function can be chosen, such as one valid for a first-order PLL [5], that is,

$$w(\varphi) = \frac{\exp(2\rho_p \cos\ \varphi)}{2\pi I_0(2\rho_p^2)}, \ |\varphi| < \pi \tag{4.78}$$

where I_0 is the modified Bessel function of the first kind and the zero-th order.
The variance of phase difference for (4.78) is

$$\sigma_\varphi^2 = \int_{-\pi}^{\pi} \frac{\exp(2\rho_p \cos\ \varphi)}{2\pi I_0(2\rho_p^2)}\varphi^2 d\varphi \tag{4.79}$$

By the Jacobi-Anger transformation [8],

$$\exp(u \cos\ \varphi) = I_0(u) + 2\sum_{n=1}^{\infty} I_n(u)\cos(n\varphi) \tag{4.80}$$

the integral in (4.79) can be solved, and the variance of phase difference becomes

$$\sigma_\varphi^2 = \frac{\pi^2}{3} + 4\sum_{n=1}^{\infty} \frac{(-1)^n I_n(2\rho_p^2)}{n^2 I_0(2\rho_p^2)} \tag{4.81}$$

The variance in (4.81) can be approximated by the term $\pi^2/3$ if the SNR is small, so the probability density function can be approximated by the Gaussian distribution function with variance $\pi^2/3$. The second term on the right side of (4.81) represents the influence of nonlinearity and is the measure of the nonlinear PLL behavior.

4.2.3.3 Influence of Light Source on Phase Variance

The variance of phase error at the output of the PLL in a heterodyne receiver can be evaluated following the analysis in Subsection 4.2.3.2. Thus the variance, σ_φ^2, of phase error can be evaluated by

$$\sigma_\varphi^2 = \int_0^\infty S(f)df \tag{4.82}$$

where $S(f)$ corresponds to the power spectral density, $\Phi(s)$, given by (4.54). At the same time, the power spectrum, $\Phi_1(s)$, in (4.54) corresponds to the FM spectrum of the IF signal. All that is valid for detection of ASK, FSK, and PSK signals.

In a DPSK system, the difference between the phase of the signal coming from the IF filter and the phase of the referent signal is given as

$$\Delta\varphi = \varphi_1(t + T_d) - \varphi_1(t) \tag{4.83}$$

where T_d denotes the digit interval. The variance of the phase error at the PLL output is now determined by (3.69) or by

$$\sigma_\varphi^2 = 4\int_0^\infty S_F(f)\sin(2\pi fT_d/f)^2df \tag{4.84}$$

(S_F is the FM noise power spectral density of the IF signal.)

The homodyne receiver includes only the OPLL, which was described in Subsection 3.3.3. The photodiode plays the role of phase detector in an OPLL, while the local laser oscillator plays the role of the VCO. Taking into account the results from Subsection 3.3.3, we can write the following equation between the Laplace transforms of phase error in the loop and the instantaneous phase of the incoming signal

$$\Phi(s) = \frac{s\Phi_1(s) - \Omega_L(s)}{s + K} \tag{4.85}$$

where K is the amplification coefficient in the OPLL, and $\Omega_L(s)$ is the Laplace transform of instantaneous frequency fluctuations of the local optical oscillator. The power spectral density of the phase error in OPLL is obtained from (4.85) and has the form

$$S(f) = \frac{S_{F,1}(f) + S_{F,L}(f)}{f^2 + (K/2\pi)^2} \tag{4.86}$$

where $S_{F,1}(f)$ and $S_{F,L}(f)$ are the FM power spectra of the optical signal source and the local optical oscillator, respectively. The phase error variance in the OPLL in the homodyne optical receiver can be determined by (4.82), where $S(f)$ takes the value given by the PLL (4.86).

Since the phase noise at the input of the PLL for both coherent receiver schemes (heterodyne and homodyne) presents FM noise of optical sources, we can find the value of phase error variance in PLL (or OPLL) by using the results for FM noise from Chapter 3. Because the FM noise spectrum of semiconductor lasers has a practically constant value in the wide range of frequencies, the total FM spectrum can be presented as the sum of two components that belong to the optical signal source in the transmitter and to the local optical oscillator in the receiver. Thus, we can write the following simple equation:

$$S_F(f) = \frac{(\delta f)_1 + (\delta f)_L}{\pi} \tag{4.87}$$

where the first and the second terms in the numerator denote the spectral linewidth of the optical signal source and the local optical oscillator, respectively. Now by using (4.53), (4.54), and (4.82) with transcriptions $\Phi(s) = S(f)$ and $\Phi_1(s) = S_F(f)$, we can easily find the phase error variance for the heterodyne detection scheme. That variance is given as

$$\sigma_\varphi^2 = \frac{(\delta f)_1 + (\delta f)_L}{2B_p} \tag{4.88}$$

where B_p denotes the equivalent bandwidth of the PLL, which in this case is $B_p = K_d K_2/2\pi$ (with the assumption that $H(s) = 1$).

The phase error variance for the DPSK detection scheme can be determined in a similar way. It follows from (4.84) and (4.87) that

$$\sigma_\varphi^2 = \frac{(\delta f)_1 + (\delta f)_L}{B_s} \tag{4.89}$$

where B_s is the modulating signal bandwidth, defined as $B_s = 1/(2\pi T_d)$.

The phase error variance for the homodyne detection scheme with application of the OPLL is obtained from (4.82) and (4.86) and has the form

$$\sigma_\varphi^2 = \frac{(\delta f)_1 + (\delta f)_L}{2B_h} \tag{4.90}$$

where B_h is the frequency bandwidth of the OPLL given as $B_h = K/2\pi$, while $S_{F,1}(f) = (\delta f)_1/\pi$ and $S_{F,L}(f) = (\delta f)_L/\pi$.

To illustrate these results, we will consider the relationships between the modulation parameters for a PSK detection scheme. Bearing in mind the results from Chapter 2 (see Figure 2.14), we can conclude that the phase error variance should be lower than 0.04 rad^2 to achieve receiver sensitivity deterioration below 3 dB. For the same linewidths of the optical signal source and the local optical oscillator, this condition leads to

$$(\delta f)_1 = (\delta f)_L < 0.02B \qquad (4.91)$$

where the bandwidth, B, takes the value $2B_p$, $2B_h$, or B_s, depending on the kind of detection (heterodyne detection of PSK signals, homodyne detection of PSK signals, or heterodyne detection of DPSK signals, respectively).

As for the influence of the resonant part of FM noise spectra, it should be mentioned that the resonant frequencies are above a few gigahertz for the injection currents, which are well above the threshold value. If the resonant frequency is much higher than the bandwidth, B, then its influence can be neglected. The authors of [9] investigated the resonant frequency influence for the case when the bandwidth, B, is considerably larger than the resonant frequency and concluded that the phase error variance is not seriously affected by the resonant effect. Hence, the resonant frequency effect in the semiconductor lasers can be neglected if the modulating bit rate is far enough from the resonant frequency, that is, the transmission bit rate should be either considerably lower or considerably higher than the resonant frequency in the laser. In such a way, the receiver sensitivity degradation due to the influence of the resonant effect in the semiconductor laser will be relatively small, which is very convenient for practical purposes.

As for the influence of the $1/f$ component in the FM noise spectrum on the phase error variance, it was shown that $1/f$ noise is the predominant component in the FM noise spectrum for high-power laser operation [10]. The influence of the $1/f$ noise component should be considered for concrete modulation methods, depending on the operating regimes of the optical signal laser and the local optical oscillator. Since the $1/f$ noise component is equal to K/f, where K is a constant, it is necessary to know the constant K that corresponds to the laser operating regime. The constant K is determined experimentally (see references [10–15] in Chapter 3). It can be assumed that the $1/f$ FM noise component imposes considerable influence on the coherent receiver sensitivity if the value of the constant K is in the region of $(1 \div 5)10^{12} \text{Hz}^2$ [10]. Such a value of the constant K can practically double the phase error variance at the end of the PLL. Possible ways to decrease the $1/f$ noise influence are either by decreasing output optical power or by introducing a redundancy in the modulating signal (by proper line coding, for example) [11].

4.3 ADVANCED VERSIONS OF COHERENT OPTICAL RECEIVERS

The basic characteristics of a coherent optical receiver and the characteristics of its key elements were considered in Sections 4.1 and 4.2. There are some serious problems related to the operation of those elements in practical coherent transmissions systems, for example, stabilization of source parameters and narrowing of the source spectral line. It is likely that the future technological realizations of relevant functional elements in both optical transmitters and optical receivers will solve most of these problems. Improved realizations of coherent optical receivers can help solve some operational problems and lead to significant improvement of receiver sensitivity. It is very convenient to realize some advanced versions of coherent optical receivers with the use of standard optical and optoelectronic elements.

The realization of a dual-photodetector coherent optical receiver offers a significant advantage over designs containing only one photodetector [12–14]. It is quite reasonable to suppose that such a realization would have large practical application. Hence, we will analyze the characteristics of dual-photodetector coherent optical receivers and point out their comparative advantages.

The sensitivity of a coherent optical receiver is determined, above all else, by the influence of shot noise, which occurs in the photodetection process of the incoming optical signals and signals from the local optical oscillator. Other factors causing degradation of receiver sensitivity are the phase noise of the optical source in the optical transmitter, the phase noise of the local optical oscillator, and the time jitter in the optical receiver. The phase noise of the local optical oscillator is, in fact, an excess optical intensity noise that is absent in DD optical receivers [15]. To suppress the influence of the excess noise, a realization with two photodiodes is employed. A dual-detector receiver has two main advantages compared with a receiver with a single photodiode. The first advantage is the possible cancellation of local oscillator excess noise; the second is more efficient use of incoming optical powers from the signal source and from the local optical oscillator. The second advantage refers to the fact that the entire incoming optical power is used in the dual-detector receiver, while some part of the incoming optical power is lost in the optical mixer (or beam splitter) if only one photodetector is employed. The design of a dual-detector optical receiver requires that total balance of the two receiving branches be achieved.

The schematic representation of the dual-detector coherent optical receiver is shown in Figure 4.5. The attenuated incoming optical signal with power P_s and carrier frequency ω_s is spatially combined with the strong optical signal from the local optical oscillator. The carrier frequency of the local oscillator signal is ω_L, and

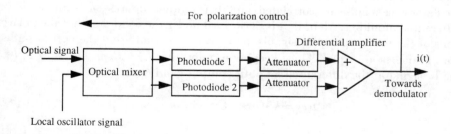

Figure 4.5 The schematic representation of the dual-detector coherent optical receiver.

its average power is P_L ($P_L \gg P_s$). The mixing of those two signals is performed in the optical mixer, which can be either a beam splitter or an optical hybrid coupler.

The optical mixing is accompanied by the phase shift, or rather the phase diversity, of π radians between two output optical signals from the optical mixer. Thus, the following equation between the output and the input optical signals at the coupler ports is valid [12]:

$$\begin{bmatrix} \mathscr{E}_1 \\ \mathscr{E}_2 \end{bmatrix} = \begin{bmatrix} \alpha/\sqrt{2} & -j\beta/\sqrt{2} \\ -j\beta/\sqrt{2} & \alpha/\sqrt{2} \end{bmatrix} \begin{bmatrix} E_s \\ E_L \end{bmatrix}, \begin{array}{l} E_s = E_s \cos \omega_s t \\ E_L = (E_L + E_{Ln})\cos \omega_L t \end{array} \tag{4.92}$$

where $\mathscr{E}_1$ and $\mathscr{E}_2$ are the values of the electrical fields of corresponding output optical signals, α and β ($\beta = 1 - \alpha$) are the port-coupling coefficients in the optical mixer, and E_s and E_L are the electrical fields of the incoming optical signal and the local oscillator optical signal, respectively. The excess noise of the local oscillator is expressed by the term E_{Ln}. The absolute values of the phases of both signals are not important for further consideration, only the phase shift between the two output-port signals.

Two optical signals from the corresponding output ports of the optical mixer are detected separately by two independent photodiodes. The obtained photocurrents from both photodiodes are subtracted by a differential amplifier, and the resulting signal is demodulated in a standard way. Thus, the current difference plays the role of a modulated electrical signal that should be demodulated by use of a PLL. The output-current signals from both photodiodes contain the IF signal carrying the sent message, mixed with the total noise. The total noise consists of shot noise, thermal noise, and excess noise of the local optical oscillator.

Since the useful signal components, contained in current signals from the photodiodes, have a mutual phase shift of π radians, the subtraction of current signals in a differential amplifier means the addition of useful signal components. The components of the shot noise in these two branches are not correlated, so their mutual suppression is not possible and they are also added. But the components of the local oscillator excess noise are correlated, and their mutual suppression is imposed in the differential amplifier. All of that is performed under the condition that the polarizations of the two output-port signals match each other and that the optical paths of both branches are equal. Thus, keeping in mind that $E_s^2 = P_s$ and $E_L^2 = P_L$ in (4.1), the IF signal at the differential amplifier output can be expressed as

$$i(t) = K4(\alpha\beta)^{1/2} E_s E_L \sin(\omega_s - \omega_L)t \tag{4.93}$$

where K is a constant corresponding to the photodiodes' sensitivity. It can be seen that the excess-noise term, E_{Ln}, disappears, which was the aim of such a receiver design.

As for the advantage of the presented scheme related to the decrease of coupling losses, it is clear that all optical power incoming to the optical hybrid (or beam splitter) is used for photodetection by either the first branch photodiode or the second branch photodiode. That is not the case in an optical receiver with only one photodiode, since the part of optical signal (about 10%) is always being lost in the beam splitter. Therefore, the realization of a coherent optical receiver with dual photodiodes increases the optical receiver sensitivity. The receiver-sensitivity improvement, compared with a single-detector case, is about 3 dB, at any rate. The improvement is higher for the case when noisy optical sources are employed, and it especially refers to the noisy local optical oscillator. Beside the dual-detector receiver, a solution with three detectors can also be applied, with a greater chance of reaching the balanced-operation regime [16].

Several factors strongly influence the operational regime of the dual-detector optical receiver. The most important of these are (a) the beam-splitting ratio α/β (or power transmission ratio) of the beam splitter or the optical hybrid coupler, and (b) the polarization states of the optical signals incoming to the optical mixer [14]. The ideal balanced dual-detector optical receiver has a beam-splitting ratio equal to 1, and every deviation of this ratio from the value of 1 degrades the performance of the receiver.

The polarization misalignment between the signals coming to the optical mixer can produce significant degradation of receiver sensitivity, so it will be necessary that the receiver be either insensitive to this polarization misalignment or capable of tracking the polarization variations of the received optical signal. The worst case that can appear is when the received signal and the local oscillator signal polarizations have opposite rotations and the major axis of the polarization state of one signal (P_s or P_L) is coincident with the minor axis of the polarization state of another signal [14].

The coherent receiver scheme (heterodyne or homodyne) is sensitive to the polarization state of the received signal, since the received signal level fluctuates with the fluctuations of the polarization states of both the incoming optical signal and the local optical oscillator signal. The fluctuations of the polarization states can be transformed into the amplitude and the phase noises as in [24] and their influence analyzed in the standard way.

Because of the great influence of polarization misalignment, a matching of the polarization states between the incoming optical signal and the local oscillator signal is required, that is, a polarization-state control scheme is indispensable in the coherent optical receiver. This problem can be completely solved by use of a polarization-preserving optical fiber over the length of the lightwave communication line. However, long-length polarization-preserving optical fibers are not at designers' disposal and probably won't be for the near future, so another solution must be considered. There are two means to eliminate the unfavorable influence of polarization fluctuations. The first one is the use of a polarization-state control device at the receiving end of the optical transmission line that matches the polarization state of the local oscillator signal with the state of the incoming optical signal [17]. The second method is the design of the polarization diversity receiver. The polarization diversity receiver is, in fact, a dual-detector receiver in which the incoming optical signal (coming from the single-mode optical line) is split into two orthogonal polarizations with defined power-splitting ratio [18]. For a splitter, a Wollaston prism can be used. After the polarization splitting, two orthogonally polarized beams are separately mixed with two local oscillator beams that have matched polarization states. The next steps are equivalent to those already described.

The two detected polarization components of signals should have equal phases, and the phase adjustment should be done in the branches before the photodetection. Under such a condition, the signals from the two branches are added coherently, and the noises from the two branches, which have no correlation with each other, are added according to power-sum rules. If the splitting coefficients in the polarization coupler are denoted by α and β ($\beta = 1 - \alpha$), then the amplitudes of the IF current signals at the input of the differential amplifier are given as

$$I_1 = 2\Re[\alpha P_L P_s]^{1/2} \text{ for the first branch} \tag{4.94}$$

$$I_2 = 2\Re[\beta P_L P_s]^{1/2} \text{ for the second branch} \tag{4.95}$$

where $\Re$ is the photodiode responsivity. It is assumed that both photodiodes have identical characteristics. The total signal power of these two current components is

$$S = \left(\frac{I_1}{\sqrt{2}} + \frac{I_2}{\sqrt{2}}\right)^{1/2} = 2\Re^2 P_L P_s(\sqrt{\alpha} + \sqrt{\beta})^2 \tag{4.96}$$

At the same time, the signal power in the one-photodiode receiver would be

$$S_1 = 4\Re^2 P_L P_s \tag{4.97}$$

Since the total-noise power stays the same in both cases, the SNR in the polarization diversity receiver will be deteriorated. The deterioration can be measured by the ratio S_1/S. In an ideal situation, when the ratio α/β in the polarization device equals unity, there will not be any deterioration of the SNR. At any rate, the deterioration in receiver sensitivity of the polarization diversity receiver is smaller than that appearing in an ordinary coherent optical receiver due to polarization misalignment.

Different schemes of polarization-diversity coherent optical receivers proposed so far [18–20] are based on similar principles as those applied in phase diversity receivers (three-detector scheme, for example). All those schemes have faster time response compared with polarization-controlling devices, because most of the feedback control process is performed in electronic circuits. Hence, there are rather complicated electronics in polarization-diversity coherent optical receivers, in order to perform all the prescribed functions.

Another solution for polarization control of the optical signal in a coherent optical receiver is employment of polarization-state control devices. Several types of such devices have been proposed, such as electromagnetic fiber squeezers, electro-optical crystals, rotatable fiber coils, and planar waveguide controllers [21–27]. All these devices feature rather complicated optics. Thus, the optimization of design of a coherent optical receiver means deciding between the application of complicated optics and the application of complicated electronics. It seems that the optical receivers based on a diversity principle will be the favored types of coherent optical receivers, but some combined schemes can also be very attractive for future realizations [28–33].

REFERENCES

[1] Kimura, T. A., and Y. Yamamoto, "Progress of Coherent Optical Fiber Communication Systems," *IEEE J. Opt. Quantum Electron.*, QE-15(1983), pp. 1–39.

[2] Personic, S. D., "Receiver Design for Digital Fiber-Optic Communication Systems—I and II," *Bell Syst. Techn. J.*, 52(1973), pp. 843–886.

[3] Van Der Ziel, A., *Noise: Sources, Characterization, Measurement*, Englewood Cliffs, N.J.: Prentice-Hall, 1970.

[4] Schwartz, M., *Information Transmission, Modulation and Noise*, 3rd ed., New York: McGraw-Hill, 1980.

[5] Goell, J. E., "Input Amplifiers for Optical PCM Repeaters," *Bell Syst. Tech. J.*, 53(1974), pp. 1771–1793.

[6] Van Der Ziel, A., *Introductory Electronics*, Englewood Cliffs, N. J.: Prentice-Hall, 1974.

[7] Viterbi, A. J., *Principles of Coherent Communications*, New York: McGraw-Hill, 1966.

[8] Korn, G., *Mathematical Handbook for Scientists and Engineers*, London: McGraw-Hill, 1961.

[9] Kikuchi, K., et al., "Degradation of Bit-Error Rate in Coherent Optical Communications Due to Spectral Spread of the Transmitter and the Local Oscillator," *IEEE/OSA J. Lightwave Technol.*, LT-2(1984), pp. 1024–1033.

[10] Kikuchi, K., and T. Okoshi, "Dependence of Semiconductor Laser Linewidth on Measurement Time: Evidence of Predominance of 1/f Noise, *Electron. Lett.*, 21(1985), pp. 1011–1012.

[11] Saito, S., et al., "S/N and Error Rate Evaluation for an Optical FSK-Heterodyne Detection System Using Semiconductor Lasers," *IEEE J. Opt. Quantum Electron.*, QE-19(1983), pp. 180–193.

[12] Singh, N., et al., "Design Considerations in Dual-Detector Receiver," *J. Opt. Commun.*, 9 (1988), pp. 150–154.

[13] Midwinter, J. E., "Optical Fiber Communications, Present and Future," *Proc. Roy. Soc. London*, 392(1984), pp. 247–277.

[14] Hodgkinson, T. D., "Receiver Analysis for Synchronous Coherent Fiber Transmission Systems," *IEEE/OSA J. of Lightwave Techn.*, LT-5(1987), pp. 573–586.

[15] Yamamoto, Y., "AM and FM Quantum Noise in Semiconductor Lasers: Parts I and II," *IEEE J. Quantum Electron.*, QE-19(1983), pp. 34–58.

[16] Davis, A. W., and S. Wright, "A Wideband Homodyne Receiver Using Phase Diversity," *IOOC-EOOC '85*, Venice, 1985, pp. 409–412.

[17] Walker, N. G., and G. R. Walker, "Polarization Control for Coherent Communications," *IEEE/OSA J. Lightwave Techn.*, LT-8(1990), pp. 438–458.

[18] Imai, T., et al., "Polarization Diversity Detection Performance of 2.5 Gb/s CPFSK Regenerators Intended for Field Use," *IEEE/OSA J. Lightwave Techn.*, LT-9(1991), pp. 761–769.

[19] Kazovsky, L., "Recent Progress in Phase and Polarization Diversity Coherent Optical Techniques," *ECOC '87 European Conf. Opt. Commun.*, Helsinki, 1987, pp. 83–90.

[20] Corvaja, R., and G. L. Pierobon, "Performance Evolution of ASK Phase-Diversity Lightwave System," *IEEE/OSA J. Lightwave Techn.*, LT-12(1994), pp. 519–523.

[21] Lefevre, H. C., "Single Mode Fiber Fractional Wave Devices and Polarization Controllers," *Electron. Lett.*, 16(1980), pp. 778–780.

[22] Ono, T., et al., "Polarization Control Method for Suppressing Polarization Mode Dispersion Influence in Optical Transmission Systems," *IEEE/OSA J. Lightwave Techn.*, LT-12(1994), pp. 891–897.

[23] Shimizu, H., et al., "Highly Practical Fiber Squeezer Polarization Controller," *IEEE/OSA J. Lightwave Techn.*, LT-9(1991), pp. 1217–1224.

[24] Okoshi, T., et al., "New Polarization Control Scheme for Optical Heterodyne Receiver Using Two Faraday Rotators," *Electron. Lett.*, 21(1985), pp. 787–788.

[25] Noda, J., et al. "Polarization Maintaining Fibers and Their Application," *IEEE/OSA Lightwave Techn.*, LT-4(1986), pp. 1071–1089.

[26] Rasleigh, S. C., "Origins and Control of Polarization Effects in Single Mode Fibers," *IEEE/OSA Lightwave Techn.*, LT-1(1983), pp. 312–331.

[27] Marrone, M. J., "Polarization Holding in Long Length Polarizing Fibers," *Electron. Lett.* 21(1985), pp. 244–245.

[28] Naito, T., et al., "Optimum System Parameters for Multigigabit CPFSK Optical Heterodyne Detection Systems," *IEEE/OSA J. Lightwave Techn.*, LT-12(1994), pp. 1835–1841.

[29] Huang, S., and J. Yan, "Power Penalty Due to Intersymbol Interference in Optical Phase Diversity FSK System," *J. Optical Commun.*, 12(1991), pp. 59–60.

[30] Felicio, D., "140 Mbit/s Optical FSK Heterodyne Dual Filter Detection System With Standard DFB Lasers," *J. Optical Commun.*, 11(1990), pp. 88–91.

[31] Iannone, E., et al., "High Speed DPSK Coherent Systems in the Presence of Chromatic Dispersion and Kerr Effect," *IEEE/OSA J. Lightwave Techn.*, LT-11(1993), pp. 1478–1485.

[32] Frederiksen, P. T., et al., "A Fast Polarization Controller for Coherent Phase Diversity Receivers," *J. Optical Commun.*, 10(1989), pp. 149–153.

[33] Tomioka, T., et al., " A Balanced Polarization Diversity Photo-Receiver With a Built in Local Laser Diode," *4th Optoelectronic Conf. OEC '92*, Chiba, Japan, 1992, pp. 84–85.

Chapter 5

Coherent Lightwave Systems

5.1 INTRODUCTION

Chapters 1 through 4 have considered in detail the characteristics of the basic functional elements of a coherent lightwave system (see Figure 1.1). Special attention was devoted to analysis of key elements in the system (light source, optical modulator, receiver amplifier) and to a description of the operation principles of basic optoelectronic schemes (photodiode front-end scheme, PLL, diversity receiver, etc.). Each of these elements or schemes determines the characteristics of the system as a whole. The performances of a coherent lightwave system are measured by the receiver sensitivity value or by the error probability in the system, under defined conditions of transmission and detection of optical signals. Since Chapter 2 was devoted to evaluation of coherent receiver sensitivity and error probability, the considerations in this chapter are in close relationship with those in Chapter 2.

The focus of the next section will be on the analysis of the characteristics of coherent lightwave systems with applied FDM technique. The possibility of realizing such systems is one of the main advantages of coherent systems over DD systems. Analysis of the characteristics of these systems is very important for their application in digital multiterminal systems and lightwave networks.

5.2 CHARACTERISTICS OF COHERENT LIGHTWAVE SYSTEMS

A lot of experimental coherent lightwave systems have been realized already [1–8], based on different theoretical methods, and using different kinds of principal elements (lasers, modulators, controllers, multiplexers).

It can be pointed out that the greater part of realized experimental systems are based on FSK and PSK (or DPSK) modulation schemes. FSK systems are, in general, more attractive than ASK systems, not only because of their higher receiver sensitivity, but because of the more convenient operational regime of the laser source in the optical transmitter (due to a simpler switching regime), as well.

The simultaneous analysis of FSK and PSK/DPSK systems shows that PSK/DPSK systems are more attractive for high bit rates, because of easier technical realization. FSK systems, despite their lower receiver sensitivity, are more attractive for lower bit rates. Hence, it is likely that FSK and PSK/DPSK coherent lightwave systems will be used most often in future telecommunications networks.

Extremely high-bit-rate coherent systems with heterodyne detection require a very large frequency bandwidth of the receiver photodiode, which might be an obstacle to their wide applications. The homodyne detection scheme offers advantages in high-bit-rate transmission systems, but under the condition that stabilization of the laser source operation is properly solved. Two practical solutions in the realizations of homodyne PSK optical receiver are the use of OPLL and the employment of diversity technique. At this moment, realizations with employed diversity technique have the advantage, but the situation could change in the future.

5.3 COHERENT SYSTEMS WITH FREQUENCY DIVISION MULTIPLEX

Coherent systems with applied frequency division multiplex technique are often called FDM coherent systems, rather than WDM (wavelength division multiplex) systems, because of the high density of the optical channels package in the multiplex spectral frame.

It is well known that the low-loss region in single-mode silica-based optical fibers extends over wavelengths from about 1.2 to 1.6 μm. That optical bandwidth is more than 30 THz. Efficient utilization of such a bandwidth requires dense packing of the optical channels, because the bandwidth of optical fiber is most easily accessed in the wavelength domain, rather than in the time domain. The application of the optical FDM technique provides dense multiplexing and the possibility of transmission of more than 1,000 optical channels simultaneously. The dense multiplexing is accompanied by the possibility of efficient demultiplexing by use of heterodyne (or homodyne) detection in optical receivers. Actually, that moment plays the main role in optical FDM systems, because it is possible to perform efficient separation of individual channels by proper frequency selection in IF stages. The wavelengths of local optical oscillators in individual coherent optical receivers are chosen to be equal to the carrier wavelengths in corresponding channels, and the filtration is made by the electrical filters. In such a way, the use of optical filters is avoided, and the finest frequency selectivity is provided. That is the main advantage in comparison with WDM optical systems, where direct detection is applied and optical filters must be used (see Section 9.3.3).

The use of the FDM technique in a coherent lightwave system means an additional increase of the system's transmission capacity. That is most important for employment of the FDM technique in long-haul transmission systems and in future broadband networks, where the use of FDM systems will not only improve the system characteristics but bring about a change in project and design philosophy. Such a claim can be illustrated with the theoretical prediction that it is possible to realize a digital network with several thousand participants where every participant will be identified by its "own" wavelength. Such a situation requires new switching principles, as well.

The principal question, related to the analysis of coherent FDM lightwave transmission systems, is what is the necessary distance between neighboring channels to prevent high intermodulation interference of those channels. Intermodulation interference is characterized by an intermodulation noise in the output spectrum of the photodiode current. Such a spectrum, consisting of some number of densely packed FDM channel spectra, is an outcome of photodetector nonlinearity.

5.3.1 Intermodulation Noise in FDM Coherent Systems

To evaluate the power of intermodulation noise in a coherent FDM system, we will consider the simplest case when photodetection is performed by only one photodiode and assume that the photodiode bandwidth is large enough. The assumption of the photodiode is necessary, because a photodiode with relatively narrow bandwidth strongly influences the output spectrum shape.

The detected signal, $s_d(t)$, at the photodiode output is proportional to the intensity of the incoming FDM optical signal, or to the square of the electrical field of the incoming FDM signal. Thus, the following relation is valid:

$$s_d(t) = K|E(t)|^2 \qquad (5.1)$$

where K is the proportional constant in a meaning of photodetector responsivity.

Equation (5.1) takes the following form in the frequency domain:

$$S_d(f) = K[E(f)*E(f)] \qquad (5.2)$$

where a convolution is denoted by the asterisk. The spectrum, $E(f)$, consists of two components: the signal spectrum, $S(f)$, and the local oscillator spectrum, $S_L(f)$ [9]. Thus, the output signal spectrum can be written in the form

$$S_d(f) = K[S(f)*S(f) + S_L(f)*S_L(f) + 2S(f)*S_L(f)] \qquad (5.3)$$

According to the aim of the analysis, it can be assumed that the spectrum, $S(f)$, consists of individual channel contributions. The first term on the right side of (5.3) is the intermodulation effect, the second term is caused by the local optical oscillator influence, and the third term is the useful part of the spectrum containing individual channels, which should be separated.

Further analysis can be performed by knowing the spectral functions $S(f)$ and $S_L(f)$. It is commonly assumed that $S(f)$ has a rectangular shape with defined width, while $S_L(f)$ is Dirac's delta function. Hence, the influences of applied line code and finite width of the local optical oscillator spectrum are neglected. These approximations were necessary to facilitate the analytic procedure. The spectral function, corresponding to the line-coded ith channel, can be represented as

$$S_i(f) = \frac{1}{4}[S_0(f + f_i) + S_0(f - f_i)], \ i = 1, 2, \ldots, N \tag{5.4}$$

where $S_o(f)$ is the power spectrum function of the line-coded signal in the ith channel before the optical modulation, and f_i is the carrier frequency of the ith channel. (The total number of channels is N.) At the same time, the spectrum of the signal from the local optical oscillator can be represented by the Lorentz function

$$S_L(f) = \frac{S_{mL}\delta f}{(f - f_L)^2 + (\delta f/2)^2} \tag{5.5}$$

where S_{mL} denotes the maximal power of the local optical oscillator, and δf represents its spectral width.

Only the spectrum of the useful signal, presented by the last term in (5.3), is of interest for further analysis. This spectrum is evaluated as

$$S_s(f) = \left[2\sum_{i=1}^{N} S_i(f)\right] * S_L(f) = \frac{1}{\pi}\int_{-\infty}^{\infty} \sum_{i=1}^{N} S_i(f - u)S_L(u)du \tag{5.6}$$

The intermodulation effect can be evaluated approximately by taking into account only one channel (for example, channel 1), and by determining the spectrum, $S_{s1}(f)$, that refers to the mutual influence of the considered channel and the local optical oscillator. The spectrum, $S_{s1}(f)$, can be evaluated as

$$S_{s1}(f) = 2S_1(f) * S_L(f) = \frac{1}{\pi}\int_{-\infty}^{\infty} S_1(f - u)S_L(u)du \tag{5.7}$$

where $S_1(f)$ is given by (5.4) for $i = 1$. Further evaluation can be performed only with the concrete shape of function $S_1(f)$. This function is given by (5.29), (5.33), and (5.35) for ASK, FSK, and PSK signals, respectively (but with transcription $f_c = f_i$). These relations, however, are not necessary for present general consideration.

It is shown in [9] that there are three general types of terms in the spectrum that are defined by (5.7). The first type contains the considered channel (channel 1) and the convolution terms between the local oscillator and the other channels. The second type corresponds to the mutual interaction between channels and presents the intermodulation noise. The third type contains the autocorrelational terms belonging to individual channels and to the local optical oscillator. It is clear that only the first type contains the information; hence, it must be filtrated by an IF filter. The IF transfer function should have the rise cosine function shape, with central frequency equal to $f_1 - f_L$ [see (5.4)].

The value of intermodulation noise can be determined by integration of (5.7) within the entire frequency range, with the exception of the IF filter frequency bandwidth. Thus, the evaluated intermodulation noise can be added to the shot noise in the optical receiver (thermal noise will be temporarily neglected). By assuming that both the shot noise and the intermodulation noise are additive processes, the SNR can be expressed as

$$SNR = \frac{P}{N_q + IN_1 + IN_2} \tag{5.8}$$

where P is the power of the IF signal, N_q is the power of shot (quantum) noise in the receiver, IN_1 is the power of the intermodulation noise caused by convolution of neighboring channels with the local optical oscillator signal, and IN_2 is the power of the intermodulation noise due to convolution between neighboring channels.

The IF signal power and the power of the noise components in (5.8) can be expressed in a common way, or as

$$P = K_1 P_L P_s \tag{5.9}$$

$$N_q = K_2 P_L \tag{5.10}$$

$$IN_1 = K_3 P_L P_s \tag{5.11}$$

$$IN_2 = K_4 P_s^2 \tag{5.12}$$

where P_s and P_L are the power of the incoming optical signal and of the local optical oscillator signal, respectively, while K_i ($i = 1, 2, 3, 4$) is the constant, depending on the photodiode responsivity and the IF filter bandwidth.

It is shown in [9] that SNR according to (5.8) should be above 16dB to reach a BER below 10^{-9}. Such a situation for DPSK signals requires a phase noise variance, σ_n^2, below 0.04 rad^2. This value of phase noise variance corresponds to the local optical oscillator spectral linewidth of about 450 kHz. It is also shown that spectral distance between neighboring channels should be larger than double the spectral linewidth of each individual channel. The channel separation has an essential role in the transmission quality. If the channel separation is below the doubled spectral width of the individual channel signal, nothing could improve the SNR to be above the value 10^{-9}. Thus, any increase in the optical power of incoming optical signal, and the local optical oscillator will not work. An optimal channel separation according to [9] is 2.2 B_s, where B_s is the spectral width of the individual channel.

It must be pointed out that this analysis refers to the heterodyne detection scheme and unbalanced (single photodetector) coherent receivers. It was shown that the application of the FDM technique is not useful for homodyne detection in unbalanced optical receivers, because of high intermodulation noise, which cannot be suppressed despite considerable increase of interchannel distance [10]. A more favorable situation will result if a balanced (double detector) receiver is applied in a homodyne detection scheme. In such a case the application of the FDM technique is quite possible in both heterodyne and homodyne detection schemes, with considerably improved transmission quality as compared with a single-photodetector case. So we will consider the balanced detection of FDM optical signals in more detail, using the basic results from [10].

5.3.2 Coherent FDM Systems with Balanced Optical Receivers

The analytical results obtained for detection of FDM optical signals in balanced coherent optical receivers are quite different from the results related to single-photodetector coherent optical receivers. Since a balanced optical receiver eliminates all the terms caused by direct detection, as well as the autocorrelational terms, the homodyne detection scheme is more attractive than the heterodyne one, because it offers denser packing of optical channels.

The schematic representation of a balanced optical receiver in Figure 4.5 will be used in the next analysis, with the assumption that just an FDM optical signal comes from the single-mode optical fiber line. The complex amplitude of a received optical signal in an FDM system is given as

$$E_s = \sum_{i=1}^{N} E_i \tag{5.13}$$

where E_i is the complex amplitude of the ith channel. This amplitude can be expressed in the form

$$E_i = \sqrt{P_s}\, m_i \exp\{ j[2\pi f_c t + 2\pi(i - M)Dt + \varphi_i\} \tag{5.14}$$

where P_s is the peak power of the optical signal, and f_c denotes the central frequency of the IF filter. Parameters m_i and φ_i refer to the possible amplitude or phase modulation of the ith channel, while D represents the channel spacing. Index M is related to the considered channel. The complex value of the amplitude of the combined optical signal reaching the first photodetector in Figure 4.5 can be expressed as

$$E_{01} = \frac{E_s + E_L}{\sqrt{2}} = \frac{1}{\sqrt{2}} \left(\sum_{i=1}^{N} E_i + E_L \right) \tag{5.15}$$

At the same time, the complex amplitude of the combined optical signal reaching the second photodetector is given as

$$E_{02} = \frac{E_s - E_L}{\sqrt{2}} = \frac{1}{\sqrt{2}} \left(\sum_{i=1}^{N} E_i - E_L \right) \tag{5.16}$$

Bearing in mind the basic equations referring to the balanced optical receiver, the current signals at the photodiode outputs in a coherent FDM optical receiver can be written as

$$i_1 = \frac{1}{2} \mathcal{R} P_L + \frac{1}{2} \mathcal{R} \sum_{i=1}^{N} \sum_{k=1}^{N} E_i E_k^* + \mathcal{R} \sqrt{P_L P_s} m_M \cos(2\pi f_c t + \varphi_M) + n_1(t) \tag{5.17}$$
$$+ \mathcal{R} \sqrt{P_L P_s} \sum_{i=1, i \neq M}^{N} m_i \cos[2\pi f_c t + 2\pi(i - M)Dt + \varphi_i]$$

$$i_2 = \frac{1}{2} \mathcal{R} P_L + \frac{1}{2} \mathcal{R} \sum_{i=1}^{N} \sum_{k=1}^{N} E_i E_k^* - \mathcal{R} \sqrt{P_L P_s} m_M \cos(2\pi f_c t + \varphi_M) + n_2(t) \tag{5.18}$$
$$- \mathcal{R} \sqrt{P_L P_s} \sum_{i=1, i \neq M}^{N} m_i \cos[2\pi f_c t + 2\pi(i - M)Dt + \varphi_i]$$

where $\mathcal{R}$ is the photodiode responsivity, and $n_1(t)$ and $n_2(t)$ are the noises in the first and second branches, respectively. The first term on the right side in (5.17) and (5.18) represents the direct-current components caused by the local optical oscillator, while the second term in each equation is the current component due to direct detection and interchannel interference. The third terms are related to the desired signal in the Mth channel, the fourth terms are noise, and the last terms are the photocurrents caused by interference between the local optical oscillator and other channels (except channel M).

The second term in (5.17) and (5.18) is the most harmful, which is in accordance with the claim that homodyne detection is not useful in single-detector detection. But the situation is quite different in this case, because the total current, i_t, is given as the difference between the currents i_1 and i_2, or

$$i_t = A\left\{ m_M \cos(2\pi f_c t + \varphi_M) + \sum_{i=1,\ i\neq M}^{N} m_i \cos[2\pi f_c t + 2\pi(i - M)Dt + \varphi_i] \right\} + n(t)$$

(5.19)

where

$$A = 2\Re\sqrt{P_L P_s}$$

(5.20)

and

$$n(t) = n_1(t) - n_2(t)$$

(5.21)

It can be seen that the first and second terms in (5.17) and (5.18) are not present in the expression for the total current, which is very convenient. Thus, the unfavorable influence of other channels will be expressed indirectly through the noise, $n(t)$, and through the contributions of the last terms in (5.17) and (5.18).

It can be assumed that the modulation bit rates and the applied modulation type are equal in all channels [10]. Thus, if the modulating signal in the considered Mth channel is $s(t) = m_M \cos(2\pi f_c t + \varphi_M)$ and the corresponding power spectral density function is $S_s(f)$, the power spectral density of the total current signal, i_t, can be expressed as

$$S_t(f) = A^2[S_s(f) + S_{int}(f)] + \nu(f)$$

(5.22)

where $\nu(f)$ is the power spectral density of noise, $n(t)$, while $S_{int}(f)$ is the power spectral density of the interference signal due to influence of other channels. This power spectral density is given as

$$S_{int}(f) = \sum_{i=1,\ i\neq M}^{N} S_i(f)$$

(5.23)

where $S_i(f)$ is the power spectral density of the signal

$$s_i(t) = m_i \cos[2\pi f_c t + 2\pi(i - M)Dt + \varphi_i]$$

(5.24)

The powers of the useful signal and interference at the IF filter output are given respectively as

$$P_{IS} = A^2 \int_{-\infty}^{\infty} |H(f)|^2 S_s(f) df$$

(5.25)

and

$$P_{int} = A^2 \int_{-\infty}^{\infty} |H(f)|^2 S_{int}(f) df \qquad (5.26)$$

where $H(f)$ denotes the transfer function of the IF filter. The parameter

$$Q = \frac{\int_{-\infty}^{\infty} |H(f)|^2 S_s(f) df}{\int_{-\infty}^{\infty} |H(f)|^2 S_{int}(f) df} \qquad (5.27)$$

presents the signal-to-total-interference ratio.

The value Q must be determined numerically, due to the complex nature of the integrals. But (5.27) can be simplified, by taking into account some practical facts referring to the coherent optical receivers. First, the frequency, f_c, should be as small as possible, because the total noise is proportional to the f_c^2 value. So it can be assumed that $f_c < D$, which means that the strongest interference noise is generated by the channel $i = M - 1$. Since $S_{M-1}(f) = S_s(f - D)$, the ratio of the signal to the strongest interference becomes

$$Q' = \frac{\int_{-\infty}^{\infty} |H(f)|^2 S_s(f) df}{\int_{-\infty}^{\infty} |H(f)|^2 S_{M-1}(f) df} = \frac{\int_{-\infty}^{\infty} |H(f)|^2 S_s(f) df}{\int_{-\infty}^{\infty} |H(f)|^2 S_s(f - D) df} \qquad (5.28)$$

The parameter Q' is very close to the parameter Q if the channel separation, D, is several times larger than the transmission digit rate, R_d. Thus, the signal-to-interference ratio can be evaluated by (5.28) for different modulation schemes in FDM systems with balanced optical receivers. These integrations are made in the frequency region corresponding to the IF filter bandwidth, B. The bandwidth, B, should be wide enough and must not influence the signal bandwidth, B_s. The signal bandwidth B_s, is most commonly defined by the width B_0 of main arcade in the spectral diagram, but a value equivalent to the double bit rate can be added to the value B_0 to obtain more precise results.

The parameter Q' can be evaluated if the value of the signal power spectrum function is known. The familiar expressions for ASK, FSK, and PSK modulation spectra can be used [12]. The power spectral density of a two-sided ASK signal is given as

$$S_s(f) = \frac{1}{16}\left[\delta(f - f_c) + \delta(f + f_c) + \frac{\sin^2 \pi T_d(f - f_c)}{\pi^2 T_d(f - f_c)} + \frac{\sin^2 \pi T_d(f + f_c)}{\pi^2 T_d(f + f_c)}\right] (5.29)$$

where T_d is the digit interval length. The width of main arcade B_0 in this case is $2R_d$ for heterodyne receivers and R_d for homodyne receivers, so the calculation of the signal-to-interference rate can be performed with the values

$$B = (2 + k)R_d, \, k = 0, \, 1, \, 2 \text{ for heterodyne detection} \tag{5.30}$$

$$B = (1 + k)R_d, \, k = 0, \, 0.5, \, 1 \text{ for homodyne detection} \tag{5.31}$$

The power spectrum function for an FSK signal depends on the modulation index, $\delta = 2f_1/R_d$, where f_1 is the deviation of the frequency, which corresponds to the half-difference between the values belonging to logical state "1" and to logical state "0". The evaluation can be performed for the orthogonal FSK signal, with $\delta = 0.5m$ ($m = 1, \, 2, \, 3, \, \ldots$), providing higher receiver sensitivity. The spectrum of such an optical signal is given by [12]

$$S_s(\beta) = \frac{T_d}{8}[\sin c(\beta - \delta/2) - (-1)^\delta \sin c(\beta + \delta/2)]^2 \tag{5.32}$$

$$+ \frac{T_d}{8}[\delta(\beta - \delta/2) + \delta(\beta + \delta/2)], \, \delta = 1, \, 2, \, 3, \, \ldots$$

and

$$S_s(\beta) = \frac{T_d}{2}[\sin c^2(\beta - \delta/2)\cos^2 \pi(\beta - \delta/2)$$

$$+ \sin c(\beta - \delta/2)\sin c(\beta + \delta/2)\cos 2\pi\beta \tag{5.33}$$

$$+ \sin c^2(\beta + \delta/2)\cos^2 \pi(\beta + \delta/2)], \, \delta = 0.5, \, 1.5, \, 2.5, \, \ldots$$

where $\beta = (f - f_c)T_d$ is the normalized frequency, and $\sin c(x) = \sin x/x$. When $\delta = 0.5$, the signal bandwidth is $B_0 = 1.5R_d$ and the parameter Q' is calculated for bandwidth

$$B = B_0 + kR_d = (1.5 + k)R_d, \, k = 0, \, 1, \, 2 \tag{5.34}$$

Finally, the power spectrum density of a PSK signal is given as [12]

$$S_s(f) = \frac{T_d}{4}\left[\frac{\sin^2 \pi T_d(f - f_c)}{\pi T_d(f - f_c)} + \frac{\sin^2 \pi T_d(f + f_c)}{\pi T_d(f + f_c)}\right] \tag{5.35}$$

with the spectrum width equal to $2R_d$ for heterodyne detection and R_d for homodyne detection. The values of the IF filter bandwidth are

$$B = (2 + k)R_d, \, k = 0, \, 1, \, 2 \text{ for heterodyne detection} \tag{5.36a}$$

$$B = (1 + k)R_d, \, k = 0, \, 1, \, 2 \text{ for homodyne detection} \tag{5.36b}$$

The evaluation of the signal-to-interference ratio, Q', was calculated in [10] for ASK, FSK, and PSK spectra. The following conclusions were made for balanced detection of optical FDM signals:

- The channel separation is proportional to the central frequency of the IF filter, that is, $D = K + 2f_c$, where K is constant depending on the modulation method and the filter bandwidth.
- The minimal channel separation is for FSK heterodyne detection with the parameters $\delta = 0.5$, $f_c = R_d$, and $B = B_0$. That separation is $D = 3.8R_d$ for $Q' = 30$ dB.
- PSK modulation requires larger channel separation than ASK modulation with the same peak power of signal. The difference, $D_{ASK} - D_{PSK}$ is about $4R_d$ and is caused by the fact that the ASK spectrum contains Dirac's function, which does not cause the interference.

These conclusions can be used in the choice of modulation scheme in an FDM coherent optical system. But more precise decision referring to the system design can be made only after considerations of other factors determining the SNR. The most important consideration is quantum noise in the FDM optical system. We will briefly analyze the influence of the quantum noise on the SNR in an FDM optical coherent system.

The single-side spectrum of the noise in a coherent optical receiver [see (4.1) to (4.11)] can be expressed as

$$\nu(f) = 2q\Re P_t + 2i_n^2, f > 0 \tag{5.37}$$

where q is the electron charge, i_n is the receiver thermal noise current caused by the influence of the photodiode, load resistance, and receiver amplifier, while P_t presents the total power of the incoming optical signal equal to

$$P_t = P_L + NP_s \tag{5.38}$$

The noise power at the IF filter output in a multichannel system with balanced receiver is

$$P_n = 2B(qRP_t + i_n^2) \tag{5.39}$$

At the same time, the noise power of an individual channel at the output of the IF filter is

$$P_{n1} = 2B[qR(P_L + P_s) + i_n^2] \tag{5.40}$$

It can be seen from (5.39) and (5.40) that the noise power in a multichannel system is larger than the noise power in a single-channel system for the factor $2B(N-1)P_s$. Thus, this factor presents a power of an additional quantum noise. The SNR at the output end of the receiver IF filter in an N-channel system can be calculated as [10]

$$Q_1 = \frac{P_{IS}}{P_n} = \frac{R^2 P_L P_s P_m}{B[qR(P_L + NP_s) + i_n^2]} \tag{5.41}$$

where P_m is the power caused by the term m_i from (5.14), having the value 0.5 for ASK modulation and the value 1 for FSK and PSK modulations. The signal-to-noise power ratio in the single-channel optical system at the output of the IF filter is higher than the value given by (5.41) and has the value

$$Q_{1c} = \frac{P_{IS1}}{P_{n1}} = \frac{R^2 P_L P_s P_m}{B[qR(P_L + P_s) + i_n^2]} = \frac{RP_s P_m}{qBY} \tag{5.42}$$

The parameter Y is defined as

$$Y = 1 + \frac{P_s}{P_L} + \frac{i_n^2}{qRP_L} \tag{5.43}$$

When the SNRs in multichannel and single-channel optical systems are identical, the corresponding BER also will be identical. The equality of these ratios can be achieved by an increase of signal power in the multichannel system. The power value that should be added is a measure of the receiver sensitivity degradation in a multichannel system and can be obtained from (5.41) and (5.42); that is,

$$\Delta P = \frac{YP_L/P_s}{YP_L/P_s - N} \tag{5.44}$$

The receiver sensitivity degradation depends on parameter Y, the number of channels, N, and the ratio of optical powers belonging to the local oscillator and the incoming optical signal in the channel. Equation (5.44) can be used in the design of FDM coherent optical lightwave systems.

The allowed receiver sensitivity degradation can be an initial parameter in (5.44) for further calculations. Another initial parameter is the number of channels, N. Hence, (5.44) must be solved to find the ratio P_L/P_s for defined initial conditions. As for parameter Y, it can be concluded that a sensitive optical receiver possesses a smaller value of parameter Y and vice versa. An ideal multichannel optical receiver possesses the value $Y = 0$. It can be assumed that low degradation in optical receiver sensitivity corresponds to the condition when $Y < 1$ dB, and that the maximal tolerable value might be $Y = 3$ dB.

To reach low receiver sensitivity degradation (below 1 dB) for a large number of optical channels (more than 200) it is necessary that the ratio P_L/P_s be above 30 dB. For the ratio P_L/P_s above 40 dB, more than 2,000 channels can be achieved with small signal degradation. It should be mentioned that the ratio P_L/P_s of 40 dB can be easily achieved in practical realization [10]. These 2,000 channels with bit rates of 1 Gb/s per channel will occupy the wavelength region of only 300 nm. It means that single mode optical fibers provide the transmission of more than 2,000 optical channels in the minimum-loss region between 1.3 μm and 1.6 μm, but the proper selection of the channels in the optical receiver should be performed.

This analysis of FDM lightwave system characteristics did not include the influence of nonlinear effects. These effects will be examined in Sections 5.3.4 and 9.3.3.

5.3.3 Application of Coherent FDM Systems in Broadband Communications Networks

IM/DD lightwave systems, with applied WDM, are characterized by a relatively small number of optical channels. That number is determined primarily by the spectral linewidth of the employed sources and the wavelength selectivity of the optical multiplexer/demultiplexer. Thus, for example, in the wavelength region around 1.3 μm, the transmission of only four optical channels is possible if standard semiconductor lasers are used as light sources. Improved versions of semiconductor lasers with considerably narrower spectral linewidth, such as DFB lasers, and the employment of tuning optical filters provide the rapid increase in channel numbers (over 40 in the spectral region around 1.3 μm). But any further increase in the number of optical channels can hardly be made because there is no possibility of efficient selection of an individual channel from a densely packed wavelength multiplexing signal (see Section 9.3.3).

The use of the FDM method in coherent lightwave systems provides the benefit of the efficient transmission of a large number of optical channels. Such an improvement is provided by the increased coherent optical receiver sensitivity in comparison with the DD receiver sensitivity and by the finest frequency selectivity in the coherent optical receiver (frequency selection is made on the electrical level). With the use of the FDM method in coherent lightwave systems, the role of some passive optical elements (such as optical couplers and optical multiplexers) is changed in relation with their role in WDM optical systems [13]. Coherent FDM systems will play an important role in future realizations of different broadband telecommunications networks. Because such an application will be dominant, we will consider some of the key aspects related to the realization of broadband communication networks that employ the FDM technique.

There are two general types of broadband networks: interactive service networks and distributive service networks. The first type allows for the mutual communication of any two participants in the network, while the latter case refers to the distribution of user services from a distributor toward participants. The main difference between these two types is in the requirements for bidirectional transmission and channel switching in interactive networks; such a requirement does not exist in distributive networks. Because of that we will analyze the main characteristics of these types of networks separately.

The principal scheme of a distributive-type lightwave network is shown in Figure 5.1. The modulated optical signals from a defined number of optical sources (or from just one source) come to the optical multiplexer or to the passive star coupler and, after that, go toward the optical fiber line. At the receiving end, there is a passive star coupler for optical signal branching toward the individual participants. The separation of the desired channel at the receiving end is performed by a coherent optical receiver with a properly adjusted IF stage. Before reaching the optical mixer at the transmitting end, the individual optical signals can be modulated either by a one-channel electrical signal or by a multiplexed electrical signal containing several electrical channels. Employment of a coherent optical receiver improves the receiver sensitivity about 15 dB in comparison with a DD optical receiver. That is very important in such an application because all the optical losses, which appeared in the optical mixer or the star coupler, are practically compensated. This fact, and the high channel selectivity in the coherent optical receivers, permits the connection of a relatively large number of participants, who have a remarkable number of information channels at their disposal.

The distributive type of lightwave networks, like the one in Figure 5.1, will have the largest application in the distribution of TV channels and broadband services. Such networks, in general, can become the interactive ones, if some of the network elements are duplicated to perform bidirectional transmission. The configuration in

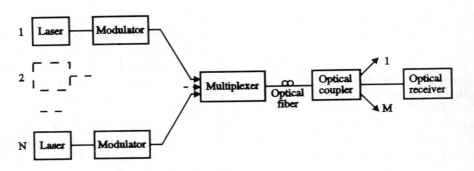

Figure 5.1 The lightwave distributive FDM network.

Figure 5.1 is the basic one, and it is possible to make more sophisticated versions and perform an optimization of the distributive lightwave networks, in accordance with the initial requirements.

The realization of interactive lightwave networks is accompanied by some problems that do not exist in distributive lightwave networks. These problems are related to the synchronization of the carrier optical frequencies that belong to the individual channels. Due to dense packing of channels in the interactive lightwave FDM networks, the laser sources can not be entirely independent and oscillate on their own frequencies, but a frequency synchronization with some referent frequency is made. That can be achieved by a frequency-locking method, which was described in Chapter 4 and which is illustrated in Figure 5.2.

As a referent oscillator, a highly stable laser, such as a He-Ne laser, can be used, which is the first stage in the frequency-stabilization scheme. The output signal from the referent laser is divided and led to a certain number of transmission stages, where the defined mutual frequency shifts are induced. These frequency shifts are relatively large (up to 10 nm), but they are defined precisely and determine the spectral width of the primary groups of optical channels. In the second transmission stage, the frequency shifts in each primary group are introduced, defining the secondary groups of the channels. The introduced frequency shift in the secondary stages are considerably smaller than the frequency shifts in the primary stage. If necessary, third stages can be introduced, but the secondary stages can provide the fine-frequency shifts, corresponding to the final spectral distance between neighboring channels. The induced frequencies are directly led from the last stage (secondary or third) to the channel laser sources. Thus, the last-stage frequencies play the role of master lasers while the channel lasers are the slaves. The master frequency can be used not only for synchronization of laser sources in the transmitter, but for synchronization of local optical oscillators as well [15].

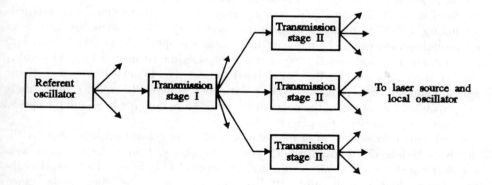

Figure 5.2 Frequency locking of a large number of lasers.

The conventional switching procedure on the electrical level can be, in general, used in the interactive lightwave networks, but some problems can arise that should be pointed out. First, either an analog or a digital technique should be used, with strongly defined channel bit rates. Second, strongly defined protocols must be applied, limiting the switching speed and complicating the whole realization. Third, realizations of switching elements for high bit rates are very expensive. All these problems led to the establishment of new switching principles.

The first of these principles includes the application of an optical switching matrix and presents an advanced realization of the scheme that has been used in conventional electronic systems. The switching matrix can be realized in different ways, but the most suitable is an integrated optoelectronic version [16]. The commutation of the optical signals is based on the same effects that are applied for external modulation of light signals (see Chapter 3). A more attractive commutation scheme is based on the application of passive optical elements and tunable coherent optical receivers.

Such a commutation scheme is illustrated in Figure 5.3 and can be performed by adjustment of the laser source frequency, the local optical oscillator frequency, or both. The commutation of optical signals can be realized either in the terminals or in a central device. The practical difficulty in realizations of such a commutation principle is the necessity that every laser source or local optical oscillator in a terminal must be locked with another one, to prevent cross-talking between individual channels. A locking principle, such as that from Figure 5.2, can hardly be realized for long distances between terminals in the network, so a centralization of the lasers can be made, as shown by the dashed line in Figure 5.3. Centralization of the lasers in the network simplifies the terminal equipment, which consists of just a modulator and a simple optoelectronic receiver [15].

The centralization of lasers achieves some additional advantages, such as easier synchronization of laser operation, possible application of a space optical commutator for distribution of the laser signals from a centralized place to the individual terminals (including optical transmitters and optical receivers), and possible application of integrated optoelectronic elements. In such a case, the optical commutator deals with nonmodulated optical signals; hence, the cross-talking effect will be considerably smaller than in schemes where modulated optical signals are commutated. If the optical commutator is sensitive to the polarization state of the optical signal, there is a possibility for control of the polarization state, since both the laser and the optical commutator are at the same place.

The main imperfection of the solution with laser centralization is that for the relatively high levels of lasers optical power is necessary. That is, the optical powers are distributed toward terminals through the optical fiber lines, which requires the considerably higher levels (up to 10 dB) of optical power than when the lasers are in the terminal equipment. Besides, three optical fibers per terminal are necessary

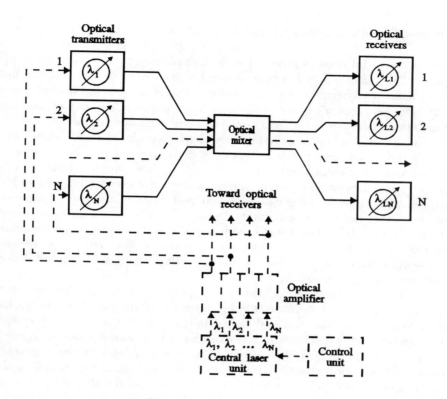

Figure 5.3 Basic commutation scheme in FDM coherent network.

(although some optimization with the application of bidirectional transmission is possible). But these problems with the centralization of lasers can be overcome by the insertion of the optical amplifier (or amplifiers) into the network, as in Figure 5.3.

The schemes considered here of broadband lightwave communications networks are just the forerunners of more perfect realizations [17, 18]. No doubt, optoelectronic devices, such as commutators, modulators, and demodulators, will be the most significant components of these systems.

Finally, it is clear that optical-domain techniques can be considered as the upgrading and complement of existing electrical time-domain techniques, such as the asynchronous transfer mode (ATM) technique and the synchronous digital hierarchy (SDH) model. With the use of coherent techniques, FDM, and photonic switching systems, telecommunications networks can gradually evolve toward broadband networks with very high capacity and wide flexibility.

5.3.4 Long-Distance Coherent Transmission Systems

Since the successful demonstration of optical amplification in erbium-doped fiber amplifiers [19], great effort has been devoted to the design of long-distance high-capacity lightwave systems. Coherent transmission over chain in-line optical amplifiers can offer high transmission capacity over long distances. The system limitations and the main limitation factors will be the subjects of analysis in this section.

System performance is generally measured through attainable transmission distances and signal bit rates and is basically determined by the allowable optical fiber loss and the allowable dispersion at the optical fiber line between the optical transmitter and the optical receiver. Optical in-line amplifiers are inserted with the main aim of improving transmission system performance in terms of attainable transmission distance for a given bit rate. Optical amplifiers enable a drastic increase in attainable optical distances because each amplifier compensates signal attenuation caused by the optical fiber loss in the preceding optical fiber section. At the same time, the transmission capacity limit caused by the dispersion basically is not changed by the optical in-line amplifiers.

Beside the optical fiber loss and fiber dispersion, the nonlinear effects in the optical fiber can be a limiting factor for system performance [20]. As we have mentioned in Chapter 2 and will analyze in detail in Chapter 6, it is generally accepted that nonlinear effects result from the concentration of high optical power in a relatively small optical fiber area. However, in a lightwave communication system with several in-line amplifiers between the transmitter and the receiver, significant optical nonlinearity can occur even at relatively low power levels because the effective path length for nonlinear effects is greatly increased. The influence of optical fiber nonlinearity on the coherent transmission design of a system is of great significance, and it is very important to determine the ultimate system capacity (measured by the product of attainable transmission distances and signal bit rates).

As already mentioned, an optical in-line transmitter is supposed to compensate for signal attenuation occurring in the preceding optical fiber section. But while amplifying the incoming optical signal, an optical amplifier generates optical noise, which is referred to as *amplified spontaneous emission* (see Chapter 8). Amplified spontaneous emission is generated by each in-line amplifier and superimposed on incoming optical signals and spontaneous emission signals caused by the preceding amplifiers in the amplifier chain. It is reasonable to assume that all amplifiers have the same characteristics in terms of signal and noise levels.

Accumulated amplified emission is the cause of beat noise in the coherent optical receiver. Several beat noise components between the optical signal, the local optical oscillator signal, and the amplified spontaneous emission noise can be observed in the coherent optical receiver [21]. Among these beat noise components, the beat noise between the local optical oscillator and the amplified spontaneous emission radiation predominates, so the other components can be neglected.

The SNR, for a heterodyne detection scheme in a coherent transmission system with cascaded in-line erbium-doped fiber amplifiers can be written as [see (8.73)]

$$R = \frac{P_s}{h\nu(G-1)N_f MB'} \tag{5.45}$$

where P_s is the amplifier output signal power, $h\nu$ is the photon energy given as a product of Planck's constant h and optical frequency ν, G is amplifier gain, N_f is the amplifier noise figure, M is the number of in-line amplifiers, and B' is the receiver filter bandwidth defined by (4.16).

Equation (5.45) can be modified as [22]

$$B_d L = \Phi L_o \frac{P_s}{h\nu(G-1)N_f} \tag{5.46}$$

where B_d is the signal bit rate, and L is the total transmission distance length, equal to $L = ML_o$, (L_o is the length between two amplifiers). Parameter Φ depends on the SNR, and is given as

$$\Phi = \frac{B_d}{RB'} \tag{5.47}$$

The left side of (5.46) is a measure of coherent transmission system capacity. It is determined by three constituents. The first parameter, Φ, is determined by coherent transmitter/receiver pair characteristics. The second parameter is the amplifier section length, while the third parameter, represented by the fraction in (5.46) is related to the SNR of the optical amplifier. Since the amplifier gain, G, depends on the length, L_o, the second and third parameters can be treated together. In fact, the fraction $L_o/(G-1)$ can be noticed by Φ_1 and expressed as

$$\Phi_1 = \frac{\ln(G)}{\gamma} \frac{1}{G-\eta} \tag{5.48}$$

where γ is the optical loss coefficient in the optical fiber per unit length, while $\eta \approx 1$ represents the total signal transmittance coefficient of the optical components in the system. These components are mainly optical isolators and optical filters inserted into the optical fiber line to improve overall characteristics through suppression of spontaneous emission effects (forward and backward propagating).

The fraction in (5.46), expressing amplifier influence, is proportional to the system capacity. As we will see in Section 8.4.2, the ratio of the erbium-doped fiber amplifier output signal power, P_s, to the amplifier noise figure, N_f, is an important measure of amplifier performance. This ratio is required to be as high as possible for an in-line optical amplifier. We can easily calculate from (5.46) that, even for ratio $P_s/N_f = 0$ dBm and for amplifier section length $L_o = 80$ km, we can expect a system capacity of $B_d L \approx 5 \cdot 10^2$ Gb/s km for both DCPSK and FSK heterodyne detection schemes [22]. This product is considerably higher for a PSK heterodyne detection scheme, while a PSK homodyne detection scheme offers additional improvement. According to these results, it seems that a transoceanic signal transmission at 10 Gb/s is achievable.

This conclusion should be proved through consideration of the influence of other relevant parameters on the coherent system capacity, which will be done in the next section. First, as we found out in Chapter 3, the spectral linewidths of the optical signal source and the local optical oscillator degrade the performance of the coherent receiver. This effect was not included in (5.46) and (5.48). The influence of the spectral linewidths of the signal light source and the local optical oscillator on the SNR can be estimated using the common approach of constituent influence on the total effect. In such a way the value of the inverse value of the resulting SNR can be expressed as the sum of two inverse-value SNR constituents:

$$1/R = 1/R_a + 1/R_\nu \tag{5.49}$$

where the first and second terms on the right side reflect the influence of in-line amplifiers and laser linewidths, respectively. So the basically negative influence of the second term can be compensated for by an additional increase in amplifier SNR, or by a decrease in the term $1/R_a$.

The spectral linewidth contribution to the total SNR will result in a change of system transmission capacity. The resulting capacity can be estimated by replacing R with R_a in (5.47) and adding the spectral linewidth contribution. This contribution was estimated in Chapter 2 for each attractive coherent detection scheme. It was found in [22] that for the ratio $P_s/N_f = 0$ dBm and for amplifier section length $L_o = 80$ km, system capacity becomes $B_d L \approx 8 \cdot 10^4$ Gb/s · km for an FSK heterodyne detection scheme. A slight decrease in amplifier spacing to $L_o = 60$ km results in a transmission capacity of $B_d L \approx 10^5$ Gb/s · km.

5.3.4.1 Influence of Fiber Dispersion on Coherent System Characteristics

Waveform distortion caused by the dispersion effect in single-mode optical fiber is a serious problem in coherent long-length lightwave transmission systems. If the effective bandwidth of each amplifier is wide enough, signal distortion due to dispersion is determined only by the influence of optical fiber dispersion. The dispersion influence on a coherent lightwave transmission system with in-line amplifiers can be determined in the same manner as for conventional point-to-point optical fiber systems. Thus, the limit due to chromatic dispersion in a single-mode optical fiber can be expressed through the product $B_d^2 L$, such as in [22], or

$$B_d^2 L = \frac{\xi c}{\lambda^2 |D(\lambda)|} \qquad (5.50)$$

where ξ is the coefficient depending on the modulation/detection scheme, c is the light velocity in the free space, λ is the optical signal wavelength, and $D(\lambda)$ is the chromatic dispersion at the signal wavelength (see Appendix J).

The coefficient ξ can be determined for the specific modulation/detection scheme, as was done in [23]. This coefficient is the highest for a DCSK scheme (about 0.738), while it has the smallest value for an FSK [either continuous phase FSK (CPFSK) or minimum shift keying (MSK)] scheme (in the range from 0.302 to 0.471). It means that an FSK scheme is unfavorable in terms of the influence of optical fiber dispersion. It is estimated in [21] that for 10-Gb/s long-distance transmission in a coherent lightwave system the optical fiber dispersion must be lower than 0.1 ps/km nm.

Beside fiber chromatic dispersion, polarization dispersion can limit coherent transmission capacity. The influence of polarization dispersion can be evaluated through [22]

$$B_d^2 L = \left(\frac{\zeta}{D_p}\right)^2 \qquad (5.51)$$

where ζ is the coefficient related to the modulation/detection format, and D_p is polarization dispersion. The coefficient ζ varies from 0.3 for a CPFSK coherent scheme to 0.4 for a PSK coherent scheme. If the polarization dispersion is lower than 0.1 ps/$\sqrt{\text{km}}$, the chromatic dispersion still remains the dispersion-limiting factor. For values above 0.2 ps/$\sqrt{\text{km}}$, polarization dispersion presents a more severe limiting factor for total transmission capacity. The statistical nature of polarization dispersion is another factor that diminishes transmission system capacity; thus polarization dispersion can be a major limiting factor in the coherent long-distance transmission system.

5.3.4.2 Influence of Nonlinearity on Coherent System Characteristics

Nonlinear effects (which will be considered in Chapter 6) can be one of the factors limiting the capacity of a coherent lightwave system. They can take an active role in the long-distance coherent lightwave system with a chain of in-line amplifiers. It is desirable to have larger amplifier output optical signals to achieve better SNRs, but that can lead to a stronger influence of the nonlinear effects on system characteristics and a higher transmission capacity will not necessarily be achieved.

If we have a chain of in-line amplifiers on the coherent lightwave transmission line, nonlinear effects will arise even if the amplifier output power is quite low. The reason for this is an extraordinary transmission line length. This is in strong contrast to conventional systems, where high-power signals from the optical transmitter cause nonlinear effects in a single-mode optical fiber. The effects caused by nonlinear refractive index dependence on the electrical field intensity, or the Kerr effect, described by (6.49), are more important in long-distance coherent lightwave systems than stimulated scattering effects.

Nonlinear dependence of the refractive index on the optical signal intensity propagating through the optical fiber causes fluctuation of the signal phase correlated with the changes in the signal intensity. Resulting optical frequency chirping, or optical spectra broadening (see Section 6.2.3), interacts with optical fiber dispersion and causes waveform distortion of the received optical signal. The intensity changes in the signal in a coherent lightwave system can be caused by the residual intensity modulation accompanying angle modulation, by frequency modulation to amplitude modulation conversion, or by amplified spontaneous emission influence generated in the optical amplifier chain. To estimate the influence of the nonlinear effects on coherent system capacity, the phase-induced noise due to optical fiber nonlinearities should be evaluated first [20].

The variance of the noise induced by nonlinearity influence is given as

$$\sigma_\varphi^2 = \frac{1}{3}\left(\frac{G-1}{G}\right)^2\left(\frac{k_2}{\gamma}\right)^2 P_s h\nu(G-\eta)N_f M^3 B' \tag{5.52}$$

where k_2 is the nonlinear coefficient given by

$$k_2 = \frac{2\pi n_2}{\lambda A} \tag{5.53}$$

where n_2 is the nonlinear Kerr coefficient, λ is the optical wavelength, and A is the effective optical fiber core area.

Equation (5.52) can be modified to give

$$B_d L^3 \cong 3\sigma_\varphi^2\left(\frac{\gamma}{k_2}\right)^2 \Phi_1(G)\frac{1}{h\nu P_s N_f} \tag{5.54}$$

where the coefficient Φ_1 has the value

$$\Phi_1(G) = \left(\frac{\ln(G)}{\alpha}\right)^3\left(\frac{G}{G-1}\right)^2\frac{1}{G-\eta} \tag{5.55}$$

The parameters M, P_s, B', N_f, γ, and η have the same meanings as in (5.45) to (5.48).

We can see from (5.54) and (5.55) that the product of the data bit rate and the third power of the transmission distance length, which is now a measure of coherent system capacity, depends on the parameters representing characteristics of the transmitter receiver pair, the optical fibers, and the in-line optical amplifiers Hence, we can say that a system's capacity is inversely proportional to the nonlinear refractive index coefficient, the optical amplifier output power, and the optical amplifier noise figure. The phase noise variance, determined by (5.52), depends on the coherent modulation/detection scheme.

To achieve the largest system capacity, an amplifier spacing of 40 to 60 km should be chosen. It is estimated in [22] that such an amplifier spacing, with noise figure of about 6 dBm and phase noise variance lower than 0.01 in (5.54), leads to the product $B_d L^3$ of $2 \cdot 10^{12}$ Gb/s $\cdot$ km^3, which corresponds to a 2-Gb/s long-distance coherent transmission system. If we compare this result with previous results obtained in this section, we can see that nonlinear effects in the optical fiber have a very strong negative influence on the performance of a long-distance coherent lightwave system.

5.3.4.3 Compensation for Influence of Dispersion and Nonlinearity in the Coherent System

The limit in coherent lightwave system capacity due to chromatic optical fiber dispersion can be alleviated by delay equalization. Such an equalizer can consist of a microstripline in the IF stages to compensate for delay distortion caused by the dispersion influence [24]. The amount of compensation is limited by the microstripline frequency response. To suppress the system capacity degradation due to dispersion influence, an external optical modulator can be used [25] instead of direct current modulation of the semiconductor laser. This method can help to reduce the spectral spread of the optical signal and contribute to the compensation dispersion influence.

Both dispersion and nonlinearity influences can be partially suppressed by applying the method suggested in [26]. In this technique, dispersion compensation by optical fibers working in anomalous and normal dispersion regions (see Figure 7.1) is utilized. In this case, Kerr-effect influence can be treated as a linear group velocity delay if the nonlinearity is small. Accordingly, optical fibers working in the anomalous dispersion region, arranged at certain intervals, can compensate for self-phase modulation effects. On the other hand, the laser chirp, which presents unwanted frequency variation of the directly modulated optical signal from a semiconductor laser, can be equalized in the optical fiber operating in the normal dispersion region. In this way, dispersion and nonlinear effects can be equalized if equalizing fibers, with a defined chromatic dispersion level, are inserted at certain intervals in the optical transmission line in the coherent lightwave long-distance system. This solution can be applied to IM/DD lightwave systems (discussed in Chapter 9), as well.

REFERENCES

[1] Waytt, R., et al., "1.52 μm PSK Heterodyne Experiment Featuring an External Cavity Diode Local Laser Oscillator," *Electron. Lett.*, 19(1983), pp. 550–552.

[2] Naito, T., et al., "Optimum System Parameter for Multigigabit CPFSK Optical Heterodyne Detection System," *IEEE/OSA J. Lightwave Techn.*, LT-12(1994), pp. 1835–1841.

[3] Iannone, E., et al "High Speed DPSK Coherent Systems in the Presence of Chromatic Dispersion and Kerr Effect," *IEEE/OSA J. Lightwave Techn.*, LT-11(1993), pp. 1478–1485.

[4] Khoe, G. D., "Coherent Multicarrier Lightwave Technology for Flexible Capacity Networks," *IEEE Commun. Mag.*, 32(1984), No. 3, pp. 24–33.

[5] Kimura, T., "Coherent Optical Fiber Transmission," *IEEE/OSA J. Lightwave Techn.*, LT-5(1987), pp. 414–428.

[6] Saito, S., et al., "An Over 2200 km Coherent Transmission Experiment at 2.5 Gb/s Using Erbium-doped Fiber In-Line Amplifiers," *IEEE/OSA J. Lightwave Techn.*, LT-9(1991), pp. 161–169.

[7] Hayashi, Y., "Implementation of a Practical Coherent Trunk System," *OFC/100C'93*, San Jose, 1993, paper ThH1, pp. 191–192.

[8] Chikama, T., et al., "Optical Heterodyne Image-Rejection Receiver for High-Density Optical Frequency Division Multiplexing System," *IEEE J. Select. Area Commun.*, 8(1990), pp. 1087–1094.

[9] Betti, S., et al., "Numerical Analysis of Intermodulation Interference in an Optical Coherent Multichannel System," *IEEE/OSA J. Lightwave Techn.*, LT-5(1987), pp. 587–591.

[10] Kazovsky, L., "Multichannel Coherent Optical Communication Systems," *IEEE/OSA J. Lightwave Techn.*, LT-5(1987), pp. 1095–1102.

[11] Personick, S. D., "Receiver Design for Digital Fiber Optic Communication System," *Bell Syst. Techn. J.*, 52(1973), pp. 843–886.

[12] Shannungam, K. S., *Digital and Analog Communications Systems*, New York: Wiley, 1979.

[13] Lucky, R. L., et al., *Principles of Data Communications*, New York: McGraw-Hill, 1968.

[14] Miller, S. E., and A. G. Chinoweth, *Optical Fiber Telecommunications*, New York: Academic Press, 1979.

[15] Stanley, I. W., et al., "The Application of Coherent Optical Techniques to Wideband Networks," *IEEE/OSA J. Lightwave Techn.*, LT-5(1987), pp. 439–451.

[16] Hinton, H. S., "Photonic Switching Systems," *European Conf. Opt. Commun.*, ECOC, Brighton, 1988, p. 239.

[17] Hill, G. R., "Multiwavelength Transport Network," *OFC/100C'93*, San Jose, 1993, paper TuJ1, pp. 43–44.

[18] *IEEE/OSA Journal of Lightwave Technology*, 11(1993), special issue on broadband optical networks.

[19] Mears, R., et al., "High Gain Rare-Earth Doped Fiber Amplifier at 1.54 μm," *OFC/IOOC '87*, Reno Nev., paper W12.

[20] Gordon, J. P., and L. F. Mollenauer, "Phase Noise in Photonic Communications Systems Using Linear Amplifiers," *Opt. Lett.*, 15(1990) pp. 1351–1352.

[21] Olson, N. A., "Lightwave Systems With Optical Amplifiers," *IEEE/OSSA J. Lightwave Techn.*, LT-7(1989), pp. 1071–1082.

[22] Saito, S., et al., "System Performance of Coherent Transmission Over Cascaded In-Line Fiber Amplifiers," *IEEE/OSA J. Lightwave Techn.*, LT-11(1993), pp. 331–341.

[23] Erlefaie, A. F., et al., "Chromatic Dispersion Limitations in Coherent Lightwave Transmission Systems," *IEEE/OSA J. Lightwave Techn.*, LT-6(1988), pp. 704–709.

[24] Tubokawa, M., et al., "Waveform Degradation Due to Dispersion Effects in an Optical CPFSK System," *IEEE/OSA J. Lightwave Techn.*, LT-8(1990), pp. 962–966.

[25] Yamazaki, S., et al., "Compensation of Chromatic Dispersion and Nonlinear Effect in High Dispersive Coherent Optical Repeater Transmission System," *IEEE/OSA J. Lightwave Techn.*, LT-11(1993), pp. 603–611.

[26] Suzukin, N., et al., "Simultaneous Compensation of Laser Chirp, Kerr Effect and Dispersion in 10 Gb/s Long Haul Transmission Systems," *IEEE/OSA J. Lightwave Techn.*, LT-11(1993), pp. 1486–1494.

Chapter 6

Nonlinear Effects in Optical Fibers

6.1 INTRODUCTION

The regime when nonlinear effects become important for transmission system design is determined by the number of parameters related to the level and the wavelength of the optical signal propagating through the optical fiber. These parameters which include spectral linewidth of the light source, number of carrier optical wavelengths (or optical channels), modulation format, optical fiber dispersion, optical fiber loss, optical fiber mode area, nonlinear refractive index coefficient in the optical fiber, optical amplifier gain, and central operating wavelength, can have considerable influence on the characteristics of the lightwave transmission system (either IM/DD or coherent).

In a nonlinear lightguide system, such as that shown in Figure 1.2, the influence of nonlinear effects might be both unfavorable and favorable. When nonlinear processes lead to additional losses, pulse broadening, and crosstalk, they are unfavorable, since they deteriorate the total transmission system capacity (increased BER, decreased total line length). On the other hand, when nonlinear processes lead to signal amplification and dispersion compensation, they are favorable, since they contribute to an increase in total transmission system capacity (increased transmission bit rate, increased total line length). We could not say that one effect is favorable and another is unfavorable in all situations. However, the classification into groups of effects with predominantly positive and predominantly negative influence can be made. Into the positive-influence group, we can include stimulated Raman scattering and soliton appearance in the optical fibers, while the negative-influence group contains stimulated Brillouin scattering, self-phase modulation, and four-foton mixing.

Section 6.2 describes these effects and their influence on nonlinear transmission system performance. A primary aim in this chapter is to describe the circumstances that favor the useful influence of nonlinear effects. Thus we will focus on nonlinear Raman and Brillouin scattering and on the soliton regime appearance, in accordance with their importance in transmission systems. Lesser attention will be devoted to a description of four-wave mixing process and self-phase modulation. During this consideration, one fact must be kept in mind: unfavorable influence of nonlinear effects should be suppressed, while favorable influence of nonlinear effects is affirmed as much as possible.

From a contemporary point of view, it is justified to regard the soliton's generation, transmission, and regeneration as the basis of the nonlinear lightwave system concept. This chapter is organized accordingly.

6.2 MAIN NONLINEAR EFFECTS IN OPTICAL FIBERS

6.2.1 Stimulated Raman and Stimulated Brillouin Scattering

Raman and Brillouin scattering processes can be explained by application of the quantum radiation theory. In that theory, these processes are considered to be two-stage reactions. In the first stage, the absorption of the incident (primarily) photons with energy hf has occurred (h is Planck's constant, and f is the frequency of optical signal). In the second stage, the emission of photons with energy hf_r is observed, accompanied by thermal or acoustic phonon emission. This emission is a result of the interaction (or rather collision) between the incident optical signal and the electronic clouds in the molecules of surrounding matter. If the secondary generated photon has a lower frequency than the incident photon, that event is called Stokes radiation. On the other hand, if the incident photon has a higher frequency than the secondary generated photon, the event is called anti-Stokes radiation. Anti-Stokes radiation, having higher frequency, corresponds to photons that have higher energy than that of the incoming light, energy that must come from the molecules. If the molecules do not have any available energy, anti-Stokes radiation vanishes. These processes are illustrated in Figure 6.1.

When both incident radiation, with frequency f, and secondary radiation, with frequency f_r, are present in matter at the same time, stimulated radiation can result. Stimulated radiation can occur only if the power of the signal with frequency f_r is higher than a necessary threshold. Stimulated radiation has the same nature as that in an ordinary laser, so the signal with frequency f plays the pump role, while the signal with frequency f_r is an induced radiation. Hence, the Raman and Brillouin processes can become the stimulated ones if the incident optical signal power is above some critical value. In that case, the scattered signal will be above the necessary threshold for stimulated radiation. In the Raman scattering process, both forward and backward traveling waves can be generated, so we have a forward or backward

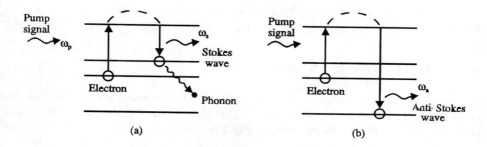

Figure 6.1 The illustration of the scattering process: (a) Stokes wave generation, (b) anti-Stokes wave generation.

Raman scattering process, respectively. At the same time, only a backward Brillouin process is possible.

In the case of forward Raman scattering, the frequency of the scattered signal is shifted to the lesser frequencies, so the scattered signal has a higher wavelength than the incident signal. This shift can lead to interchannel interference in a WDM system. On the other hand, backward Raman or Brillouin scattering can decrease the level of the forward traveling wave, which carries the useful information. That occurs because of the energy exchange between the information forward wave and the scattered backward wave.

It is clear, however, that stimulated scattering can be used for the amplification of optical signals. In such a case, the pump signal must have enough energy to induce stimulated scattering. The scattered signal frequency must be very close to the frequency of the information-carrying signal, which is introduced independently on the pump. Hence, the scattered signal contributes to an increase in the information signal, that is, it amplifies the information signal. Both the Raman and the Brillouin amplifiers based on stimulated scattering can be realized, but the Raman amplifiers are more efficient in practical applications.

To analyze the stimulated scattering process in an optical fiber, we must start from the differential equations that describe the propagation of scattered and pump waves. These equations are solved under the assumption that the pump signal decays exponentially with the optical fiber length. The main aim of solving the equations is the evaluation of the pump-power level that is necessary for the stimulated scattering appearance. That critical level for forward scattering can be evaluated under the assumption that the scattered signal must be smaller than the incident pump power at any point along the line. The necessary critical level for backward scattering is found assuming that the scattered signal at the input end of the optical fiber must be lower than the incident optical signal at the same end.

The forward and backward scattering processes will be analyzed separately. The backward-process analysis will include both Raman and Brillouin scattering.

6.2.1.1 Stimulated Forward Scattering

Forward stimulated scattering can be considered in an optical fiber with effective cross-sectional area A and total length L. The loss constant of the optical fiber will be denoted by γ. All parameters referring to the pump signal will have the subscript p, while the parameters referring to the scattered radiation will have the subscript r. The pump power, P_p, having the density $S_p \simeq P_p/A$, decays exponentially with the coordinate z, or

$$P_p(z) = P_p(0)\exp(-\gamma_p z) \tag{6.1}$$

Assuming that optical fiber is an active medium for stimulated scattering, it is easy to form the differential equation for a forward traveling wave or Stokes wave having the frequency f_r. Such an equation for the optical power has the form [1]

$$\left(\frac{d}{dz} + \gamma_r\right)P_r(z) = \alpha S_p(z)P_r(z) \tag{6.2}$$

where α is the amplification constant depending on the frequency and on the cross-sectional area of the fiber core.

It can be assumed that a decrease in pump power is not affected by the nonlinear interaction but only by the optical fiber loss, so the following equation is valid:

$$\left(\frac{d}{dz} + \gamma_r\right)P_r(z) = \alpha P_r(z)S_p(0)\exp(-\gamma_p z) \tag{6.3}$$

where $S_p(0)$ is the input pump power density. By solving this equation, the optical power $P_r(z)$ stands for

$$P_r(z) = P_r(0)\exp\left\{-\gamma_r z + \frac{\alpha S_p(0)}{\gamma_p}[1 - \exp(-\gamma_p z)]\right\} \tag{6.4}$$

Now, taking that $z = L$ and $\gamma_p L \gg 1$, the power of the nonlinear scattered signal at the output end of the optical fiber becomes

$$P_r(L) \simeq P_r(0)\exp\left[-\gamma_r L + \frac{\alpha S_p(0)}{\gamma_p}\right] \tag{6.5}$$

It follows from (6.1) and (6.5) that the total amplification of the signal with the frequency f_r, defined by the coefficient

$$\alpha_t = \frac{\alpha S_p(0)}{\gamma_p} \tag{6.6}$$

is equal to the gain produced by the pump for the distance $z = L_{ef} = 1/\gamma_p$.

The power of the scattered wave in (6.5) is evaluated under the assumption that a Stokes wave arises at the point $z = 0$. But the situation is quite different in reality, and there is not the incident Stokes wave; all the power of the Stokes wave at the output end of the optical fiber is the result of the amplification of a spontaneously scattered signal along the total line length. To explain this process, the Stokes wave should be expressed by the sum of components belonging to individual transverse modes. At the same time, the total signal (rather, the number of photons) per mode should be evaluated.

The differential equation for the number of photons in a Stokes mode has the form [1]

$$\left(\frac{d}{dz} + \gamma_r\right) N_r = \alpha S_p(z)(N_r + 1) \tag{6.7}$$

The term $\alpha S_p(z)$ in (6.7) takes into account the spontaneous emission influence. The following equation is valid for the spontaneous emission of photons:

$$\frac{d}{dz} N_{r,sp} = \alpha S_p(z) \tag{6.8}$$

By neglecting the spontaneous emission, the solution to (6.7) becomes

$$\frac{N_r(z_2)}{N_r(z_1)} = \exp\left[\gamma_r(z_1 - z_2) + \int_{z1}^{z2} \alpha S_p(z) dz\right] \tag{6.9}$$

It can be assumed that the photons, which arise due to a spontaneous process on the incremental length, $\alpha S_p(z) dz$, will be amplified according to (6.9). The sum of the number of photons in the spontaneous emission mode and the number of incident photons is

$$N_r(z) = \int_0^z \alpha S_p(x) \exp[\gamma_r(x - z) + \int_x^z \alpha S_p(y) dy] dx \tag{6.10}$$

$$+ N_r(0) \exp[-\gamma_r z + \int_0^z \alpha S_p(y) dy]$$

The number $N_r(z)$ from (6.10) is the solution to (6.7). Since the pump power density decreases exponentially with length z, that is,

$$S_p(z) = S_p(0)\exp(-\gamma_p z) \tag{6.11}$$

the number of photons can be expressed as

$$N_r(z) = \int_0^z dx \, \alpha S_p(0)\exp\left\{-\gamma_p x + \gamma_r(x - z) + \frac{\alpha S_p(0)}{\gamma_p}[\exp(-\gamma_p x) - \exp(-\gamma_p z)]\right\} \tag{6.12}$$

$$+ \, N_r(0)\exp\left\{-\gamma_r z + \frac{\alpha S_p(0)}{\gamma_p}[1 - \exp(-\gamma_p z)]\right\}$$

Since there is no incident Stokes wave, the number of incident photons is zero, or $N_r(0) = 0$. Thus, the final number of photons at the point $z = L$ becomes

$$N_r(L) = \exp(-\gamma_r L)\int_0^L \alpha S_p(0)\exp\left\{\frac{\alpha S_p(0)}{\gamma_p}[\exp(-\gamma_p x) - \exp(-\gamma_p L)]\right\}dx \tag{6.13}$$

Equation (6.13) was obtained under the assumption that $\gamma_r = \gamma_p$. Since $\gamma_p L \gg 1$ for systems used in practice, the last term under the integral in (6.13) can be neglected. If the stimulated emission plays the main role, it can be assumed that $\alpha S_p(0)/\gamma_p \gg 1$ and $\gamma_p x < 1$. By using the functional expansion $\exp(-\gamma_p x) \simeq 1 - \gamma_p x$, (6.13) takes the form

$$N_r(L) \simeq \exp\left[-\gamma_r L + \frac{\alpha S_p(0)}{\gamma_p}\right]\int_0^L \alpha S_p(0)\exp[-\alpha S_p(0)x]dx \simeq \left[-\gamma_r L + \frac{\alpha S_p(0)}{\gamma_p}\right] \tag{6.14}$$

It can be seen that the same result will be obtained from (6.12) if $N_r(0) = 1$ and spontaneous emission is neglected. Hence, the main conclusion is that the result of amplification of total spontaneous radiation is equivalent to the fictitious injection of a single photon at the point $z = 0$. This conclusion is essential for the next considerations.

Now, the total power of a Stokes wave can be expressed by the sum of the mode powers, that is,

$$P_r(L) = \sum^M \int hf \exp\left[-\gamma_r L + \frac{S_p(0)}{\gamma_p}\alpha(f)\right]df \tag{6.15}$$

The total number of modes is designated by M, while the amplification coefficient is expressed as a function of frequency f. If this function has the Lorentz shape with full width δf and maximum value α_0, the previous equation becomes

$$P_r(L) = \left\{ \sum^{M} \int hf_r \exp\left[-\gamma_r L + \frac{S_p(0)\alpha_0}{\gamma_p}\right] \right\} B_{ef} \tag{6.16}$$

where B_{ef} is the effective bandwidth of amplification. The effective bandwidth is calculated as

$$B_{ef} = \frac{\sqrt{\pi}}{2} \frac{\delta f}{[S_p(0)\alpha_0/\gamma_p]^{1/2}} \tag{6.17}$$

The effective input Stokes power is

$$P_{r,ef}(0) = hf_r B_{ef} M \tag{6.18}$$

The stimulated scattering process will not play an important role if the Stokes power, $P_r(L)$, is smaller than the pump signal power at the point $z = L$, that is,

$$P_{r,ef}(0)\exp\left[-\gamma_r L + \frac{S_p(0)\alpha_0}{\gamma_0}\right] < P_p(0)\exp(-\gamma_p L) \tag{6.19}$$

Considering the inequality in (6.19) as an equality, the critical pump power, P_{cr}, necessary for the beginning of nonlinear Raman scattering, can be found. The critical power is the smallest in the single-mode optical fibers, because $M = 1$. That power can easily be found under the assumption that $\gamma_r = \gamma_p$, so the following relation is valid:

$$\frac{\sqrt{\pi}}{2} hf_r \left(\frac{\alpha_0}{A\gamma_p}\right) \delta f = \left(\frac{\alpha_0 P_{cr}}{A\gamma_p}\right)^{3/2} \exp\left(-\frac{\alpha_0 P_{cr}}{A\gamma_p}\right) \tag{6.20}$$

The approximate solution to (6.20) is

$$P_{cr} = \frac{16A\gamma_p}{\alpha_0} \tag{6.21a}$$

which is often used in the practical calculations.

The Raman amplification coefficient α depends on the kind of material and must be determined experimentally. The typical dependence of coefficient α on the wavelength shift, $\delta\lambda = (\lambda_p - \lambda_s)$, for silica-based optical fibers is shown in Figure 6.2. The amplification coefficient refers to the pump wavelength $\lambda_p = 1.3$ μm. Figure 6.2 is obtained by fitting the experimental results from [2] and [3]. The dopants, as boron or germanium, in silica-based optical fibers will not appreciably change the functional shape of the coefficient α. The curve from Figure 6.2 is, in fact, related to a polarization-maintaining optical fiber. In a conventional optical fiber that does not maintain polarization, the amplification coefficient is approximately only half the value expressed in Figure 6.2 ($\alpha_0 \simeq 5 \cdot 10^{-12}$ cm/W). As for wavelength shift in stimulated Raman scattering process, it is within the range 35 to 55 nm and cannot be exactly evaluated.

Accordingly, the values of the critical optical power given by (6.20) and (6.21) are related to the polarization-stable single-mode optical fiber. In a conventional optical fiber, however, these values will be doubled.

The critical optical power can be calculated for a typical single-mode optical fiber. Taking, for example, that $\gamma_p = 0.3$ dB/km, $\gamma_p = 1.5$ μm, $A = a^2\pi = 50$ μm^2 (a is the core diameter), it is obtained that $P_{cr} \simeq 1.7$ W. This value is much higher than a value coupled into the optical fiber in practical optical fiber communication systems. But the nonlinear Raman amplification can appear at much lower optical powers if an optical WDM is applied. In such a system, the energy from one channel may be transformed into the energy of another channel with higher wavelength. As a criterion for the critical optical power in such a case, it can be assumed that the amplification of the signal in one channel by the signal from another channel must be less than 1% [2]. Thus, the critical optical power in the WDM system is given by

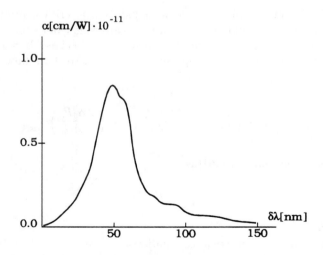

Figure 6.2 The shape of Raman amplification coefficient for silica-based single-mode optical fiber.

$$P_{cr} = \frac{0.01 A \gamma_p}{\alpha_0} \tag{6.21b}$$

This critical power has the value of about 1 mW for a typical single-mode optical fiber. Hence, the nonlinear Raman scattering can considerably influence the characteristics of WDM systems, which will be discussed in Chapter 9.

6.2.1.2 Stimulated Backward Scattering

Stimulated forward Raman scattering has three parameters. In this process, the incident pump photons induce the Stokes wave photons and high-frequency optical phonons. During that process, phase matching between incident and new generated particles (photons and phonons) is attained. Such phase matching occurs in backward Raman scattering process, as well. But the Brillouin amplification process can occur only in a backward direction because the phase-matching requirements for Brillouin scattering prohibit forward scattering. That is because the three-parameter process in this case, besides the incident and the Stokes photons, includes the acoustic phonons. Hence, acoustic phonons rather than high-frequency optical phonons arise in the Brillouin scattering process.

Analysis of backward scattering can be made in a manner similar to that of the forward scattering, but keep in mind the backward direction of the scattered wave propagation. Such an approach can be applied to either Raman or Brillouin scattering. Hence, the incident pump wave travels in $+z$ direction between the points $z = 0$ and $z = L$, while the Stokes wave propagates in $-z$ direction. The number of the scattered-wave photons can be determined from the differential equation

$$\left[\frac{d}{dz} - \gamma_r \right] N_r = -\alpha S_p(z)(N_r + 1) \tag{6.22}$$

By neglecting the spontaneous-emission term on the right side of (6.22), its solution takes the form

$$\frac{N_r(z_2)}{N_r(z_1)} = \exp\left[\gamma_r(z_2 - z_1) + \int_{z1}^{z2} \alpha S_p(z) dz \right] \tag{6.23}$$

But including the spontaneous emission, as well as $N_r(L)$ photons injected at the point $z = L$, this solution becomes [1]

$$N_r(z) = \int_z^L \alpha S_p(x) \exp[\gamma_r(z - x) + \int_z^x \alpha S_p(y) dy] dx \tag{6.24}$$

$$+ N_r(L) \exp[\gamma_r(z - L) + \int_z^L \alpha S_p(y) dy]$$

If we take that $S_p(z) = S_p(0)\exp(-\gamma_p z)$, (6.24) becomes

$$N_r(z) = \int_z^L \alpha S_p(0)\exp\left\{\gamma_r z - (\gamma_r + \gamma_p)x + \frac{\alpha S_p(0)}{\gamma_p}[\exp(-\gamma_p z) - \exp(-\gamma_p x)]\right\}dx \quad (6.25)$$

$$+ N_r(L)\exp\left\{\gamma_r(z - L) + \frac{\alpha S_p(0)}{\gamma_p}[\exp(-\gamma_p z) - \exp(-\gamma_p L)]\right\}$$

With no injected Stokes wave, (6.25) at the input point $z = 0$ becomes

$$N_r(z) = \int_0^L \alpha S_p(0)\exp\left\{-(\gamma_r + \gamma_p)x + \frac{\alpha S_p(0)}{\gamma_p}[1 - \exp(-\gamma_p x)]\right\}dx \quad (6.26)$$

Now, assuming that $\gamma_p = \gamma_r = \gamma$ and under the condition that $\gamma L \gg 1$, the number of scattered photons can be expressed as

$$N_r(0) = \frac{\exp[\alpha S_p(0)/\gamma_p]}{\alpha S_p(0)/\gamma_p} \quad (6.27)$$

The net gain, G, experienced by a Stokes wave is, in fact, the difference

$$G = \alpha S_p(z) - \gamma_r \quad (6.28)$$

If the pump signal decreases exponentially, the net gain becomes zero when it is

$$\alpha S_p(0)\exp(-\gamma_p z_0) = \gamma_r \quad (6.29)$$

or

$$\gamma_p z_0 = \ln[\alpha S_p(0)/\gamma_r] \quad (6.30)$$

Assuming that only one photon is injected at the point $z_0 = L$ and neglecting the spontaneous emission, (6.25) can be written in the condensed form

$$N_r(0) = \frac{\exp[\alpha S_p(0)/\gamma_p - 1]}{\alpha S_p(0)/\gamma_p} \quad (6.31)$$

Thus, under the assumption that $\alpha S_p(0)/\gamma_p \gg 1$, (6.31) takes the form of (6.27). Hence, the net effect of amplification of the spontaneous-emission signal is equal to the injection of a single fictitious photon per mode at the point z_0, which is determined by (6.30).

The total gain of the Stokes wave injected at point z_0 is defined by the ratio $P_r(0)/P_r(z_0)$, that is,

$$G_t = \frac{P_r(0)}{P_r(z_0)} = \exp\left\{-\gamma_r z_0 + [\alpha S_p(0)/\gamma_p][1 - \exp(-\gamma_p z_0)]\right\} \quad (6.32)$$

The total power of the Stokes backward scattered wave at the input point of the optical fiber can be found by summation of all spontaneous emission contributions, but now multiplied by gain G_t. Since it has already been shown that this summation is approximately equal to the injection of only one Stokes photon per mode at the equipose point, z_0, the effective gain of a single fictitious photon is equal to the number $N_r(0)$, that is,

$$G_{ef} = \frac{\exp[\alpha(f)S_p(0)/\gamma_p - 1]}{\alpha(f)S_p(0)/\gamma_p} \quad (6.33)$$

The total scattered Stokes power is

$$P_r(0) = \sum^{M}\int hf G_{ef}(f)df \quad (6.34)$$

The effective bandwidth of the Stokes backward signal and the effective input Stokes power in the backward direction are the same as for the case of forward scattering and are given by (6.17) and (6.18), respectively. But there is, in fact, a difference between backward Raman scattering and Brillouin scattering. The phonons that appear in Brillouin scattering are thermally activated, so the effective input Stokes power given by (6.18) should be multiplied by the parameter ϑ [1]

$$\vartheta = \frac{1}{\dfrac{kT}{hf_a} + 1} \simeq \frac{kT}{hf_a} \quad (6.35)$$

where k is the Boltzmann constant, T is the absolute temperature, and f_a is the frequency of the acoustic phonon [4]. The parameter ϑ is equal to the average phonon number (typically 100 to 200) and depends on the pump wavelength and the sound velocity.

The critical power for backward scattering is defined as the input power at the point $z = 0$, which is equal to the backward scattered stimulated Stokes power at the same point. This value can be easily found for the single-mode optical fiber. Thus, the critical power, P'_{cr}, for the stimulated Raman backward scattering is found from

$$\frac{\sqrt{\pi}}{2}(hf_r)\frac{\alpha_0(\delta f)_R}{\gamma_p A} = \left(\frac{\alpha_0 P'_{cr}}{\gamma_p A}\right)^{5/2} \exp\left(-\frac{\alpha_0 P'_{cr}}{\gamma_p A}\right) \tag{6.36}$$

while the critical power, P''_{cr}, for the stimulated Brillouin scattering is found from

$$\frac{\sqrt{\pi}}{2}\left(\frac{f_r}{f_a}\right)(kT)\frac{\alpha_0(\delta f)_B}{\gamma_p A} = \left(\frac{\alpha_0 P''_{cr}}{\gamma_p A}\right)^{5/2} \exp\left(-\frac{\alpha_0 P''_{cr}}{\gamma_p A}\right) \tag{6.37}$$

The approximate solutions of (6.36) and (6.37) are

$$P'_{cr} = \frac{20 A \gamma_p}{\alpha_0} \tag{6.38}$$

and

$$P''_{cr} = \frac{21 A \gamma_p}{\alpha_0} \tag{6.39}$$

respectively.

Equations (6.21) and (6.38) give the critical powers for stimulated forward and stimulated backward Raman scattering, respectively. It can be seen that the critical power for backward scattering is higher than the critical power for forward scattering. The critical power for stimulated Brillouin scattering will be equal to the critical power for stimulated backward Raman scattering only if the corresponding amplification coefficients are equal. Equations (6.21), (6.38), and (6.39) differ only in the value of the multiplication numerical factor. In each case, this factor is approximately the natural logarithm of the gain necessary to bring the spontaneous emission to the level of the input pump power [1].

The peak value of the amplification coefficient for the stimulated Brillouin scattering is [5]

$$\alpha_0 = \frac{(2\pi^2 f_r f_a M_e)}{c^2 \gamma_a} \tag{6.40}$$

where f_r and f_a are the frequencies of the Stokes photon and the acoustic phonon, respectively; γ_a is the loss coefficient of the acoustic wave; c is the light velocity; and M_e is the elasto-optic parameter. The parameter M_e is given by

$$M_e = n^6 p^2 / \rho v_a^3 \tag{6.41}$$

where n is the refractive index, p is the elasto-optic constant, ρ is the material density, and v_a is the acoustic wave velocity. It was found [4–5] that the stimulated Brillouin amplification coefficient in single-mode optical fiber is over two orders of magnitude larger than the amplification coefficient for stimulated Raman scattering, that is, $\alpha_0 \approx 4 \cdot 10^{-9}$ cm/W.

To find the critical power for the stimulated Brillouin scattering, all parameters from (6.40) and (6.41) should be known, some to be found by experiment. According to the results from [2–5], the critical power for a typical single-mode optical fiber is about 35 mW, for the pump wavelength near 1 μm. Another very important parameter, referring to the stimulated Brillouin scattering, is the bandwidth of the scattered signal. This bandwidth can be determined experimentally, and its typical value is about 50 MHz [4]. (We can see from Figure 6.2 that the bandwidth of a Raman signal is considerably wider.)

As for the frequency shift in the stimulated Brillouin process, the scattered light is shifted down to a lower frequency by the amount $\Delta f = 2nv_a/\lambda$, where n is the refractive index, and v_a is the acoustic wave velocity in the fiber. At $\lambda = 1.55 \, \mu$m, this shift is about 11 GHz for silica glass fiber [6].

The critical powers for forward and backward stimulated scattering were determined under the assumption that the pump signal is monochromatic, having wavelength λ_p. But when the pump signal has a finite spectral width, (6.21), (6.38), and (6.39) are valid until the spectral width of the pump signal is lower than the spectral width of the scattered signal. The spectral width δf_R of the stimulated Raman signal is high enough to consider that (6.21) and (6.38) are valid for the laser sources. On the other hand, due to the small spectral width of the stimulated Brillouin signal, (6.39) is valid only for monomode lasers. The critical powers that we have discussed were determined for continual wave-pump operation. However, (6.21), (6.38), and (6.39) also are valid for the pulse pump operation but only under two conditions: (1) the pulse width of the pump must be so short that its envelope bandwidth is higher than the spontaneous emission linewidth, and (2) the difference between the pump wave group velocity and the Stokes wave group velocity has to be greater than separate pulse width multiplied by the factor $1/\gamma_p$. In case (2) the interaction length, or active length for stimulated scattering, is shortened. In the case of backward stimulated scattering, the interactive length has the boundary value

$$L_{ef} = v_g \Delta t/2 \qquad (6.42)$$

where v_g denotes the group velocity, while Δt is the pulse duration. The boundary value, L_{ef}, is related to the pulse pump operation. Such a limit is caused by the opposite directions of the propagations of the pump and the Stokes wave. The critical power of an individual pulse exceeds the values defined by (6.38) and (6.39). When the pump signal consists of very narrow pulses and when the inverse value of the pulse width is higher than the bandwidth of the scattered signal, the transient nature of the scattering process must be taken into account. In such a case, the critical pump power increases, but there is a possibility of successive amplification of the backward Stokes wave. This possibility is valid especially for the Brillouin scattering process. Hence, stimulated Brillouin scattering can be also used in the optical amplifier design (which is the positive side of stimulated Brillouin scattering application; the negative side is the possible increase of forward traveling signal attenuation). Because both stimulated Raman and stimulated Brillouin scattering have practical meaning only in single-mode transmission, we will not consider these processes in the multimode optical fibers. It should just be mentioned that the critical power of the pump in such a case must be much higher.

But another moment may have practical meaning in the stimulated scattering process. It is clear from (6.21), (6.38), and (6.39) that a small reduction of the input pump power greatly reduces the level of the scattered signal. Thus, a 1-dB reduction of the input pump power reduces the scattered signal level by nearly 20 dB. That means that only a slight decrease of the input signal is necessary for eventual suppression of unfavorable influence of the scattered signal on the transmission system characteristics. On the contrary, a significant increase of the scattered signal needs only a small increase of the pump signal above the critical value. The practical aspects of these conclusions will be discussed in Chapter 9.

6.2.2 Four-Photon Mixing Process

Four-photon mixing (or three-wave mixing) is a nonlinear parametric process that can be observed in the optical fibers simultaneously with the Raman scattering, but the stimulated Raman scattering is not the condition for the four-photon mixing process. The shift between the frequencies of the generated signal and the pump signal in the four-photon mixing process is considerably larger than that for stimulated Raman scattering. Simultaneously, the amplification coefficient for four-photon mixing is generally higher than that for stimulated Raman scattering, so we can expect that the output power of the four-photon process wave will be higher than the Raman output power. But it was experimentally observed that this fact is true only for the optical fiber lengths shorter than some characteristic length, L_c [7], while for lengths longer than L_c the stimulated Raman signal dominates.

The four-photon process in optical fiber is illustrated in Figure 6.3. During the propagation of the pump signal through the optical fiber, the absorption of a couple of photons is occurring, and a couple of new, secondary photons are being generated. One of the new photons has a higher, the other a lower frequency than that of the pump. The photons with lower frequency belong to the Stokes wave, while the photons with higher frequency belong to the anti-Stokes wave. The main difference compared with Raman process is that the four-photon process needs strong phase matching, while the Raman process can appear even without any phase matching.

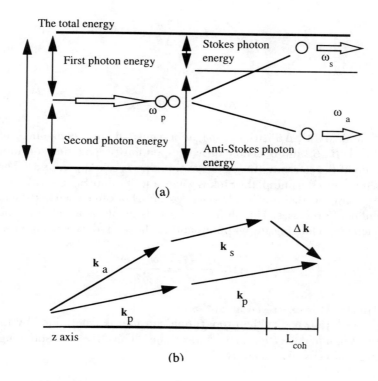

Figure 6.3 Illustration of the four-photon mixing process.

The bandwidth of the signal generated in the four-photon mixing process depends on the pump signal intensity. At the same time, the frequency shift of the maximum gain depends on pump intensity, as well. These effects can be directly related to an intensity-dependent refractive index. Next, we will explain four-photon mixing in more detail and discuss the conditions for the phase matching. The discussion is directly related to multimode optical fibers, but obtained results can be applied to single-mode fibers, as well.

The phase matching between the primary and the secondary photons in the four-photon mixing process cannot be achieved easily, because of the dispersion effects in the optical fiber. Thus, the ideal matching does not occur. Because of that, the wavelength of the secondary generated wave is slightly different from the frequency of the corresponding free-running wave. Due to nonideal phase matching, the resulting phase slippage will increase with the length of the optical fiber. After some coherence length, L_{coh} [7], the phase of the secondary generated wave will be for π radians out of the free-running wave phase, and the energy will begin to come back to the pump wave. The coherence length is

$$L_{coh} = \frac{2\pi}{\Delta\beta} \tag{6.43}$$

where

$$\Delta\beta = 2\beta_p - \beta_s - \beta_a \tag{6.44}$$

In (6.43) and (6.44), $\Delta\beta$ is the propagation constant (or z component of the wave vector), while β_p, β_s, and β_a denote the corresponding propagation constants (defined by the ratio of the refractive index to wavenumber k; see Appendix J). The subscript p, s, and a refer to the pump, the Stokes wave, and the anti-Stokes wave, respectively. Another parameter, the so-called frequency shift, Ω, is often used in the consideration of all nonlinear processes. This shift is defined as the normalized difference between the frequencies belonging to the pump and to the secondary generated wave, or

$$\Omega = \frac{\omega_p - \omega_s}{2\pi c} = \frac{\omega_a - \omega_p}{2\pi c} \tag{6.45}$$

The parameter Ω is expressed in cm^{-1} units.

It is well known that the propagation constant is determined by the material and the waveguide dispersion in an optical fiber. Thus, the corresponding contributions can be directly expressed by

$$\Delta\beta = \Delta\beta_m + \Delta\beta_w \tag{6.46}$$

where $\Delta\beta$ is defined by (6.44), and $\Delta\beta_m$ and $\Delta\beta_w$ refer to the material and the waveguide dispersion, respectively. It was shown that $\Delta\beta_m$ and $\Delta\beta_w$ depend on parameter Ω [6] and that $\Delta\beta_m(\Omega)$ can be approximated by

$$\Delta\beta_m(\Omega) = 2\pi\lambda D_m(\lambda)\Omega^2 \tag{6.47}$$

where $D_m(\lambda)$ is the well-known material dispersion function, given by (J.4) (see Appendix J). The material dispersion function can be rewritten as

$$D_m(\lambda) = \lambda^2 \frac{d^2n}{d\lambda^2} \tag{6.48}$$

Phase matching in the four-photon mixing process can be achieved with different combinations of the pump signal modes in the multimode optical fiber. The four-photon process can be observed in the single-mode optical fiber, as well, but single-mode, polarization-preserving optical fiber must be used. In such a case, two polarization states in the fiber play the roles of the separate modes. Phase matching in the multimode (rather, dual mode) and single-mode optical fibers is shown in Figure 6.4. Figure 6.4(a) illustrates the dual-mode optical fiber. Parameters b and V from Figure 6.4 have their usual meaning of normalized propagation constant and normalized frequency, respectively [8]. It can be observed that all pump photons belong to the fundamental LP_{01} mode, while the Stokes and anti-Stokes photons belong to the LP_{11} mode. Phase matching will occur if the value of $\Delta\beta$ [according to (6.46)] takes the value zero. The value of $\Delta\beta$ becomes zero for the relatively large value of the material dispersion component because of the large contribution of the waveguide component. The large value of the material dispersion requires a large value of frequency shift, as provided by the scheme in Figure 6.4(a). That shift is typically in the range of 1,000 to 2,000 cm^{-1}, while the coherence length lies in the range of 10 to 50 cm [9].

The four-photon mixing process in a single-mode optical fiber is illustrated in Figure 6.4(b). The phase matching occurs due to splitting of two polarizations of the fundamental mode (LP_{1x} and LP_{1y}) [10]. The parametric amplification in this case is smaller than that in the multimode fiber, because a part of the mode (but not all of it) is involved in the process. The corresponding frequency shift is smaller and lies in the range of 100 to 1,000 cm^{-1}, while the coherence length is in the range of 1m to 2m.

The most commonly observed four-photon mixing process will happen if the pump is divided between two fundamental modes (LP_{01} and LP_{11}). Then the Stokes

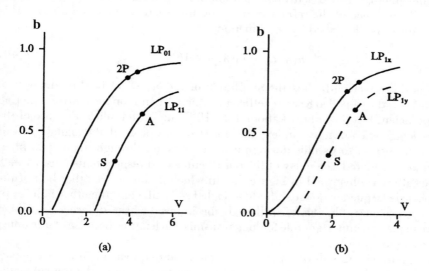

(a)   (b)

Figure 6.4 The representative combinations of the modes for phasematching: (a) multimode fiber, (b) single mode fiber. (*After*: [7], © 1982 IEEE, reprinted with permission.)

wave belongs to the LP_{11} mode, while the anti-Stokes wave belongs to the LP_{01} mode. In such a case, the coherence length is around 10m, while the frequency shift is in the range of 100 to 500 cm^{-1}. But in this range of frequency shift the stimulated Raman process is often present. Thus, we have the mixing of two nonlinear processes, which is a complicating factor not only for analysis but for any practical realization of optical amplifiers, as well.

It is interesting to discuss the case in which the coherence length is relatively large and the frequency shift is very small. For example, it might be that the coherence length is about 2 km and the frequency shift is about 1 cm^{-1}. The coherence length increases with a decrease in the dispersion parameter $D(\lambda)$, so the coherence length can be even higher in the minimum-dispersion region. Because of the large coherence length and the small frequency shift, there is not necessarily another phase matching in the process. This process can lead to mixing with a stimulated Brillouin wave and include the self-phase modulation, as well. Thus, it is clear that analysis in such a case is difficult.

In conclusion, we can say that the four-photon mixing process is not practical for use in communications systems; it can be one of the factors limiting transmission capacity in multichannel lightwave systems.

6.2.3 Self-Phase Modulation

The nonlinear effect called self-phase modulation occurs as a consequence of the nonlinear dependence of the refractive index on the electrical field amplitude. This dependence can be described by the function

$$n(\lambda, E) = n_1(\lambda) + n_2|E|^2 \tag{6.49}$$

where λ is the wavelength, E is the amplitude of the electrical field of the optical signal, and n_2 is the so-called Kerr's coefficient, which depends on the kind of material. Kerr's coefficient has the value of about $1.2 \cdot 10^{-18}$ cm^2/V^2 for silica. Because of the intensity-dependent refractive index, the phase of the peak of the signal pulse is delayed with respect to its tails during pulse propagation through the optical fiber. That can be considered as a delay of the pulse center with respect to its tails, as well.

This effect is illustrated in Figure 6.5, in which it can be seen that half of the output pulse is frequency down-shifted (so-called red shift) and the other half is up-shifted in frequency (blue shift). Although the induced phase difference is rather small, it can play an important role for long transmission lengths, because of unwanted signal spectrum spreading.

The phase changes during pulse propagation are equivalent to the frequency modulation, or "chirp" effect. The chirp effect can arise even in the optical signal generation process. The frequency variation, $\Delta f(t)$, of a directly modulated single-mode semiconductor laser can be represented by [11]

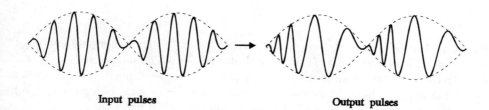

Input pulses   Output pulses

Figure 6.5 Illustration of the self-phase modulation effect.

$$\Delta f(t) = -\frac{\alpha}{4\pi}\left[\frac{d}{dt}\log P(t) + \chi P(t)\right] \qquad (6.50)$$

where $P(t)$ is the output optical power, χ is the so-called damping coefficient, and α is the linewidth enhancement factor defined by (3.73).

The self-modulation effect can be compensated for by the dispersion effect, which will cause the elimination of negative consequences of both effects. This compensation induces the soliton regime, which will be described in detail in Chapter 7.

The phase shift along the optical fiber length, L, induced by the intensity-dependent refractive index, is

$$\Delta\Phi = \frac{2\pi\Delta n}{\lambda}L \qquad (6.51)$$

where $\Delta n = n(\lambda, E) - n_1(\lambda)$. The phase shift changes in time proportionally to the pulse intensity. This change leads to the spectral broadening, $\delta\omega$, which is given by

$$\delta\omega = -\frac{d(\Delta\Phi)}{dt} \qquad (6.52)$$

This spectral broadening is related to the spectral width, $\Delta\omega$, of the optical signal. For example, for Gaussian pulses we have that [12]

$$\delta\omega = 0.86\Delta\omega\Delta\Phi_m \qquad (6.53)$$

where $\Delta\Phi_m$ is the maximum phase shift in radians. This shift is proportional to the signal peak power, the coefficient n_2, the effective length, L, of the optical fiber, and the inverse value, $1/A$, of the cross-sectional area of optical fiber. The significant spectral broadening is observed for $\Delta\Phi_m \geq 2$. The peak power corresponding to the value $\Delta\Phi_m = 2$ is [12]

$$P_m = \frac{\lambda A}{120L}, \text{ W} \tag{6.54}$$

The optical peak power, P_m, can be calculated for the typical single-mode optical fiber; thus, it is obtained that P_m is around 30 mW for single-mode fiber.

To estimate the extent of the pulse spreading and pulse distortion at the output end of the optical fiber, the definition equations for the propagation constant and the group velocity should be used. The group velocity, v_g, is given by

$$v_g = \frac{d\omega}{d\beta} \tag{6.55}$$

where β is the propagation constant. It is well known that the pulse spreading and the pulse distortion are caused by the variation of the group velocity (or the propagation constant) with optical frequency ω. Since the propagation constant, β, depends on the electrical field intensity, E, as well, it is clear that this dependence is another cause of the pulse distortion. Taking into account the functional dependence $\beta(\omega, E)$, the following functional expansion can be made:

$$\beta(\omega, E) = \beta_0 + \beta'(\omega - \omega_0) + \frac{1}{2}\beta''(\omega - \omega_0)^2 + \frac{1}{6}\beta'''(\omega - \omega_0)^3 + \frac{2\pi}{\lambda}n_2|E|^2 \tag{6.56}$$

where ω_0 is the carrier frequency, $\beta_0 = \beta(\omega_0)$, and the primes denote the derivative of a function with ω as an argument. The term β_0 causes the phase shift of the optical wave, the term β' is the cause of the pulse delay, while the terms β'' and β''' cause the pulse broadening and the pulse distortion, respectively. The nonlinear term $n_2|E|^2$ leads to the self-phase modulation. If the influence of the nonlinear term is compensated for by the β''-term influence, the soliton regime in optical fiber can be achieved. Using (6.55) and (6.66), the following expansion can be made:

$$\frac{\tau_g}{L} = \frac{1}{v_g} = \beta' + \beta''(\omega - \omega_0) + \ldots \tag{6.57}$$

where τ_g is the pulse group delay τ_g over a length, L (see Appendix J).

Now we will examine the influence of the nonlinear term on propagation characteristics. The electrical field, $E(z, t)$, of the optical signal, which propagates through the optical fiber, is commonly expressed in the form [8]

$$E(z, t) = \text{Re}\{u(z, t)\exp[j(\omega_0 t - \beta_0 z)]\} \tag{6.58}$$

where $u(z, t)$ is the pulse envelope function. With a quantum mechanics approach, we can make the following replacements:

$$\beta - \beta_0 = j\frac{\partial}{\partial z} \tag{6.59a}$$

$$\omega - \omega_0 = -j\frac{\partial}{\partial t} \tag{6.59b}$$

Now, applying (6.56) to the pulse envelope function, we obtain the following nonlinear differential equation:

$$j\frac{\partial u}{\partial z} = -j\beta'\frac{\partial u}{\partial t} - \frac{1}{2}\beta''\frac{\partial^2 u}{\partial t^2} + \frac{j}{6}\beta'''\frac{\partial^3 u}{\partial t^3} + \frac{2\pi n_2}{\lambda}|u|^2 u \tag{6.60}$$

This equation, the well-known Schrodinger-type equation, is the base equation for the consideration of nonlinear pulse transmission in optical fibers. This form of the nonlinear equation does not include the influence of the optical fiber loss. The exact form of the nonlinear equation will be given in Chapter 7. Equation (6.60) can help us make the main conclusion referring to the soliton regime. Thus, the second and the last terms on the right side of (6.60) contribute to the pulse broadening during its propagation. But the nature of the broadening is quite different. The second term is related to the dispersion effects, while the last term is related to the nonlinear self-phase modulation effect. The self-phase modulation can be the main factor in the pulse broadening only if the transmission is made at the wavelength from the zero dispersion region (around 1.3 μm for silica-based single-mode optical fibers).

6.3 THE SOLITON REGIME IN PULSE PROPAGATION

Notice that the second and the last terms on the right side of (6.60) have opposite signs. Hence, we can suppose that full compensation between these two terms is possible under specific operating conditions. These conditions concern the optical signal power level, the shape of the signal envelope, the operating wavelength, the optical fiber loss, and so on. Under these conditions, the pulse will preserve its input shape and will be stable over the entire transmission length. Such pulses are called *solitons*, while the conditions for their propagation are the *soliton regime*. Soliton generation and transmission open a new way toward extremely high bit-rate transmission over long lengths, and herald a new era in the development of lightwave telecommunications [13–15].

Chapter 7 will be devoted to the consideration of soliton generation and soliton transmission and to the main aspects of their employment in communication systems, while Chapter 9 considers the characteristics of nonlinear lightwave systems, in which soliton generation and transmission are achieved. Special attention in Chapter 9 will be paid to soliton-pulse reshaping (and regeneration) over long-length lines [16–20].

REFERENCES

[1] Smith, R. G., "Optical Power Handling Capacity of Low Loss Optical Fibers as Determined by Stimulated Raman and Brillouin Scattering," *Appl. Optics*, 11(1972), pp. 2489–2494.

[2] Stolen, R., "Nonlinearity in Fiber Transmission," *IRE Proc.*, 68(1980), pp. 1232–1236.

[3] Stolen, R. G., and E. P. Ippen, "Raman Gain in Glass Optical Waveguides," *Appl. Phys. Lett.*, 22(1973), pp. 276–278.

[4] Ippen, E. P., and R. G. Stolen, "Stimulated Brillouin Scattering in Optical Fibers," *Appl. Phys. Lett.*, 21(1972), pp. 539–541.

[5] Heinman, D., et al., "Brillouin Scattering Measurement in Optical Glasses," *Phys. Rev.*, 819(1979), pp. 6583–6592.

[6] Chraplyvy, A. R., "Limitations on Lightwave Communications Imposed by Optical Fiber Nonlinearites," *IEEE/OSA J. Lightwave Techn.*, LT-8(1990), pp. 1548–1557.

[7] Stolen, R. H., and J. E. Bjorkholm, "Parametric Amplification and Frequency Conversion in Optical Fibers," *IEEE J. Quantum Electron.*, QE-18(1982), pp. 1062–1071.

[8] Keiser, G., *Optical Fiber Communications*, Tokyo: McGraw-Hill, 1983.

[9] Stolen, R. H., et al., "Phase Matched Three Wave Mixing in Silica Fiber Optical Waveguides," *Appl. Phys. Lett.*, 24(1974), pp. 308–310.

[10] Stolen, R. H., et al., "Phase Matching in Birefringent Fibers," *Opt. Lett.*, 6(1981), pp. 213–215.

[11] Koch, T. L., and J. E. Bowers, "Nature of Wavelength Chirping in Directly Modulated Semiconductor Lasers," *IEE Electron. Letters*, 20(1984), pp. 1038–1039.

[12] Stolen, R. H., et al., "Self-Phase Modulation In Silica Optical Fibers," *Phys. Rev.*, A17(1978), pp. 1448–1454.

[13] Stolen, R. H., et al., "Optical Kerr Effect In Glass Waveguide," *Appl. Phys. Lett.* 33(1973), pp. 294–296.

[14] Doran, N. J., and K. J. Blow, "Solitons in Optical Communications," *IEEE J. Quantum Electron.*, QE-19(1983), pp. 1883–1888.

[15] Hasegawa, A., and Y. Kodama, "Signal Transmission by Optical Soliton in Monomode Fibre," *IRE Proc.*, 69(1981), pp. 1145–1150.

[16] Rottwit, K. G., et al., "Long Distance Transmission Through Distributed Erbium Doped Fibers," *IEEE/OSA J. Lightwave Techn.*, LT-11(1994), pp. 2105–2114.

[17] Nakashima, T., et al., "Configuration of the Optical Transmission Line Using Stimulated Raman Scattering for Signal Light Amplification," *IEEE/OSA Lightwave Techn.*, LT-4(1986), pp. 569–573.

[18] Molemauer, L., et al., "Long Distance Soliton Propagation Using Lumped Amplifiers and Dispersion Shifted Fiber," *IEEE/OSA J. Lightwave Techn.*, LT-9(1991), pp. 194–197.

[19] Kodama, Y., and A. Hasegawa, " Generation of Asymptotically Stable Optical Solitons and Suppression of the Gordon-Haus Effect," *Opt. Lett.*, 17(1992), pp. 31–33.

[20] Chandrakumar, V., et al., "Combination of In-Line Filtering and Receiver Dispersion Compensation for an Optimized Soliton Transmission," *IEEE/OSA J. Lightwave Techn.*, LT-12(1994), pp. 1047–1051.

Soliton Transmission in Optical Fibers

7.1 INTRODUCTION

The main features of the solitons in optical fibers were pointed out in Section 6.2. Detailed consideration of the soliton regime includes definition of conditions for soliton generation and soliton propagation and analysis of soliton behavior in optical fibers. The losses in optical fibers are the main factor deteriorating the soliton regime. Besides, too small an initial distance between two neighboring solitons can cause deterioration of the soliton regime. Too small an initial distance can also result in the mutual interaction between the neighboring soliton pulses and the appearance of intensive time jitter. The initial distance is defined as an initial condition, which is changed during soliton propagation, under the influence of optical fiber loss.

Analysis of soliton transmission refers to single-mode optical fibers, above all else. The conditions for soliton generation and transmission can be easily achieved in a single-mode optical fiber. A relatively low optical power is needed for dispersion-effects compensation by nonlinear self-phase modulation effects at the prescribed wavelength (usually in the vicinity of 1.5 μm).

This chapter analyzes the general characteristics of the soliton regime first and then examines the influence of the most important factors (loss, third-order dispersion, fiber inhomogeneity).

7.2 SOLITON GENERATION IN OPTICAL FIBERS

Instead of the propagation constant, β, which is only the imaginary part of the complex propagation constant $\hat{\gamma} = \gamma + j\beta$, just the constant $\hat{\gamma}$ can be used in (6.63), so it becomes [1]

$$j\left(\frac{\partial u}{\partial z} + \gamma u + \beta_0' \frac{\partial u}{\partial t}\right) - \frac{1}{2}\beta_0'' \frac{\partial^2 u}{\partial t^2} - \frac{j}{6}\beta''' \frac{\partial^3 u}{\partial t^3} + \frac{1}{2}\frac{\omega_0 n_2}{c}|u|^2 u = 0 \qquad (7.1)$$

The value $(\gamma)^{-1}$ is the e-folding distance of the electric field amplitude due to optical fiber loss (e is the base of natural logarithm). In the absence of any loss, dispersion, or nonlinearity, (7.1) would have the solution in the form [2]

$$u = u(t - z/v_g) \qquad (7.2a)$$

According to this solution, the following coordinate transformation can be made:

$$t' = t - z/v_g \qquad (7.2b)$$

(The group velocity, v_g, is equal to β''). On the other hand, it is suitable to normalize the values of z, t', and u by Hasegawa's [1] way, that is,

$$\xi = 10^{-9}z/\lambda \qquad (7.3a)$$

$$\tau = \frac{10^{-4.5}t'}{(-\beta''\lambda)^{1/2}} \qquad (7.3b)$$

$$q = 10^{-4.5}(\pi m_2)^{1/2}u \qquad (7.3c)$$

Equation (7.1), which determines the normalized electrical field envelope change, now becomes

$$j\frac{\partial q}{\partial \xi} + \frac{1}{2}\frac{\partial^2 q}{\partial \tau^2} + |q|^2 q = -j\Gamma q + j\overline{\beta}\frac{\partial^3 q}{\partial \tau^3} \qquad (7.4)$$

where

$$\Gamma = 10^9 \lambda \gamma \qquad (7.5)$$

$$\overline{\beta} = \frac{1}{6}\frac{\beta_0'''}{\beta_0''}\frac{10^{-4.5}}{(-\lambda\beta_0'')^{1/2}} \qquad (7.6)$$

These normalizations are made such that the main parameters ξ, τ, and q take the values of unity order. That means that $\xi = 1$ for $\lambda = 1.3$ μm and $z \simeq 1.3$ km, while $\tau = 1$ means $t = 1.7$ ps. At the same time, the value $q = 1$ corresponds to the electrical field amplitude $u = 2 \cdot 10^6$ V/m. It can be easily checked that, under such normalizations, the left side of (7.4) has the value about unity for a picosecond pulse transmission through a kilometer optical fiber length. The argument τ in (7.3b) has the real value in the wavelength region where $\beta_0'' < 0$. This region is the so-called *anomalous dispersion region* (while the wavelength region where $\beta_0'' > 0$ is called the *normal dispersion region*). The regions of anomalous and normal dispersions are presented in Figure 7.1. It is clear that only transmission in the anomalous dispersion region provides mutual compensation between dispersion and nonlinear effects.

It is useful to look at the quantities on the right side of (7.4). For an ideal optical fiber with loss coefficient $\gamma = 0.2$ dB/km at $\lambda = 1.3$ μm, we have that $\Gamma \approx 0.03$. The value of parameter $\overline{\beta}$ depends on the optical fiber design and the operating wavelength, λ. For standard single-mode optical fibers, the parameter $\overline{\beta}$ has a value of about -0.002 in the anomalous dispersion region.

Since all the terms on the right side of (7.4) are much smaller than those on the left side, the right-side terms can be treated as perturbations to a homogeneous equation. So it is quite justified first to consider (7.4) without the right-side terms and then include their influence, as in [1].

The equation

$$j\frac{\partial q}{\partial \xi} + \frac{1}{2}\frac{\partial^2 q}{\partial \tau^2} + |q|^2 q = 0 \qquad (7.7)$$

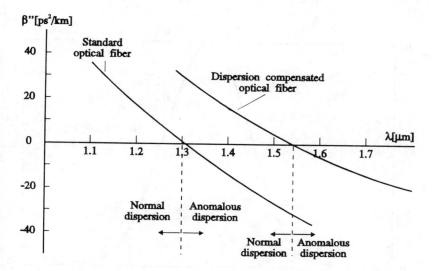

Figure 7.1 Regions of normal and anomalous dispersion in standard and dispersion compensated optical fiber.

is a homogeneous, Schroedinger-type nonlinear equation, giving a soliton solution, as is well known from theory [2]. In theory there exists the formalism of solving the Schroedinger equation for the known input function $q(\tau, 0)$, which is based on the application of the inverse-scattering theory. The general solution of (7.7) can be expressed by N solitons and by some number of continuous modes. The number N is determined by the area

$$A = \int_{-\infty}^{\infty} |q(\tau, 0)| d\tau \qquad (7.8)$$

The stationary solution for $|q|$ is

$$q(\tau, \xi) = q_0 \, \text{sech}(q_0\tau)\exp(jq_0^2\xi/2) \qquad (7.9)$$

The amplitude of this solution has the hyperbolic secant type shape, as shown in Figure 7.2. Such an impulse is sometimes called "the bright soliton," because the soliton pulse brings energy, or "brightness."

The solution of (7.7) is also the function

$$\tilde{q}(\tau, \xi) = q(\tau - \frac{\xi}{\zeta}, \xi)\exp\left[j\left(\frac{\tau}{\xi} - \frac{1}{2}\frac{\xi}{\zeta^2}\right)\right] \qquad (7.10)$$

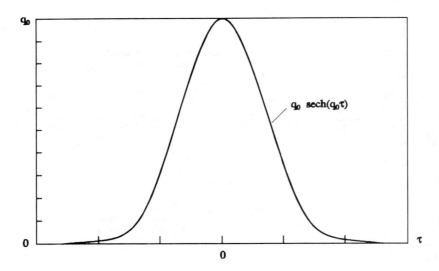

Figure 7.2 The shape of bright soliton pulse in optical fiber.

This function indicates that the frequency modulation at the point $\xi = 0$ can generate soliton propagation at a velocity different from that without modulation. This fact can be very important for future research of combined coherent-nonlinear lightwave systems.

The solution of (7.7) exists only in the anomalous dispersion region, because τ is a real variable. The value of $2t_0$ that satisfies the condition $q_0 t_0 = 1$ is the pulse width at which the pulse height becomes equal to $\text{sech}(1) = 0.65$. The relationship between the peak value, u_0, of the electrical field and the pulse width, t_0, can be obtained using (7.3b) and (7.3c), that is,

$$u_0 = \frac{1}{t_0} \left(-\frac{\lambda \beta_0''}{\pi n_2} \right)^{1/2} \tag{7.11}$$

It is more suitable to find the peak value of the input optical power, P_0, which produces solitons. Thus, (7.11) takes the form

$$P_0 = \frac{v_g u_0^2 \epsilon_0 S n_0^2}{2} \tag{7.12}$$

where S is the cross-sectional area of the fiber core, and ϵ_0 is the dielectric constant in the vacuum ($\epsilon_0 = 8.854 \cdot 10^{-12}$ F/m).

Hence, the power, P_0, depends on both the dispersion, β_0'', and the wavelength, λ. If $\beta_0'' = 0$, P_0 becomes zero as well, which is a quite logical consequence. But that is only the theoretical conclusion; practically speaking, it means that the optical power has the minimal value at the zero dispersion point. Thus, for the typical values of a single-mode optical fiber, it follows from (7.11) and (7.12) that [1]

$$(2t_0)^2 P_0 \approx 1.6 \text{ W} \cdot \text{ps}^2 \tag{7.13}$$

Now we have the practical question of how to reach the soliton regime. In accordance with the results of the inverse-scattering theory, every real pulse with the area $A \simeq \sqrt{1.6} \text{W}^{1/2}$ ps ($\pm 50\%$) can generate one soliton. Thus, the soliton number induced by the input pulse

$$q(\tau, 0) = a q_0 \text{ sech}(q_0 \tau) \tag{7.14}$$

will depend on the coefficient a. The soliton number will be the closest integer that satisfies the condition

$$a + \frac{1}{2} \geq N \tag{7.15}$$

For $0.5 \leq a \leq 1.5$, the integer N takes the value 1, so only one soliton is generated. Hence, only one soliton is to be generated for the hyperbolic secant shape of the input pulse having 1-ps width and 1.6-W peak power. This conclusion is valid in the absence of loss at the wavelength $\lambda = 1.3 \ \mu m$. Only one soliton will be generated if the input pulse has power in the region of 0.4 to 3.6 W, as well.

If the soliton amplitude coefficient, a, in (7.14) is

$$a = 1 + \alpha, \ |\alpha| < \frac{1}{2} \qquad (7.16)$$

the corresponding asymptotic value of the pulse height for an infinity length is

$$a_\infty = 1 + 2\alpha \qquad (7.17)$$

Thus, the input pulse having a width equal to 1 ps and the peak power $P_0 = 3.6$ W gives $\alpha = 0.5$ and $a_\infty = 2$. The asymptotical power of the pulse is proportional to a_∞^2 and will be 6.4 W. In such a case, the total width of the output pulse is lower than the input pulse width, that is, $(1 + 2\alpha)^{-1}q_0^{-1} = 0.5$ ps. So there is the possibility for pulse compression, if the parameter α is lower than 0.5. For $\alpha \geq 0.5$, the total soliton number in the output pulse increases, which is not suitable for digital signal transmission.

It is important to point out that there is some oscillation period, L_0, necessary for the input shape to reach the soliton shape. During that period, the pulse shape is being changed, and it loses some amount of energy. The oscillation period can be found by taking the nonlinear term of (7.1), or

$$\frac{1}{L_0} \simeq \frac{1}{2} \frac{\omega_0 n_2}{c} |u|^2 \qquad (7.18)$$

The soliton period indicates the soliton's behavior along the transmission path. During this time, the periods of compression and dilation are exchanging, and after that the pulse takes the initial shape.

7.2.1 Influence of Third-Order Dispersion and Loss on Soliton Behavior

To find the solution of inhomogeneous (7.4), we can apply the perturbation method [1, 3]. The effects of the third-order dispersion term and the loss term can be considered separately. We will consider the influence of the third-order dispersion term first. In this case, (7.4) takes the form

$$j\frac{\partial q}{\partial \xi} + \frac{1}{2}\frac{\partial^2 q}{\partial \tau^2} + |q|^2 q = j\,\overline{\beta}\,\frac{\partial^3 q}{\partial \tau^3} \qquad (7.19)$$

The function $q(\tau, \xi)$ can be expressed as a power of $\overline{\beta}$, or

$$q = q_{(0)} + \overline{\beta} q_{(1)} + \dots \tag{7.20}$$

where $q_{(0)}(\tau, \xi)$ is the solution of (7.7) given by (7.9). Applying the expansion (7.20) in (7.19), it becomes

$$j\frac{\partial q_{(1)}}{\partial \xi} + \frac{1}{2}\frac{\partial^2 q_{(1)}}{\partial \tau^2} + 2|q_{(0)}|^2 q_{(1)} + q_{(0)}^2 q_{(1)}^* = j\frac{\partial^3 q_{(0)}}{\partial \tau^3} \tag{7.21}$$

where the asterisk denotes the conjugate-complex value. The term on the right side contains the resonant solution, that is, a solution of (7.7), which generates a secular solution to (7.21). Consequently, the right-side term should be eliminated by the following scheme [1]:

$$j\frac{\partial q_{(0)}}{\partial \xi} + \frac{1}{2}\frac{\partial^2 q_{(0)}}{\partial \tau^2} + |q_{(0)}|^2 q_{(0)} + \overline{\beta}\left[j\frac{\partial q_{(1)}}{\partial \xi} + \frac{1}{2}\frac{\partial^2 q_{(1)}}{\partial \tau^2} \right.$$
$$\left. + 2|q_{(0)}|^2 q_{(1)} + q_{(0)}^2 q_{(1)}^* \right] = j\overline{\beta}\frac{\partial^3 q_{(0)}}{\partial \tau^3} + \theta(\overline{\beta}^2) \tag{7.22}$$

Notice that (7.22) is equivalent to (7.21). Now we can add the term

$$Q = \delta\omega q_{(0)} + j\delta\lambda \frac{\partial q_{(0)}}{\partial \tau} + \overline{\beta}\left(\delta\omega q_{(1)} + j\delta\lambda \frac{\partial q_{(1)}}{\partial \tau} \right) \tag{7.23}$$

to both sides of (7.22) and after that expand $\delta\omega$ and $d\lambda$ as follows:

$$\delta\omega = \overline{\beta}\omega_1 + \overline{\beta}^2\omega_2 + \dots \tag{7.24}$$

$$\delta\lambda = \overline{\beta}\lambda_1 + \overline{\beta}^2\lambda_2 + \dots \tag{7.25}$$

Finally, the coefficients ω_1, ω_2, λ_1, and λ_2 should be determined so as to cancel out the resonant term on the right side of (7.21). Then, the following equations are obtained:

$$j\frac{\partial q_{(0)}}{\partial \xi} + \frac{1}{2}\frac{\partial^2 q_{(0)}}{\partial \tau^2} + |q_{(0)}|^2 q_{(0)} + \delta\omega q_{(0)} + j\delta\lambda \frac{\partial q_{(0)}}{\partial \tau} = 0 \tag{7.26}$$

$$j\frac{\partial q_{(1)}}{\partial \xi} + \frac{1}{2}\frac{\partial^2 q_{(1)}}{\partial \tau^2} + 2|q_{(0)}|^2 q_{(1)} + |q_{(0)}|^2 q_{(1)}^*$$
$$+ \delta\omega q_{(1)} + j\delta\lambda \frac{\partial q_{(1)}}{\partial \tau} = \frac{\partial^3 q_{(0)}}{\partial \tau^3} + \omega_1 q_{(0)} + j\lambda_1 \frac{\partial q_{(0)}}{\partial \tau} \tag{7.27}$$

It follows from (7.26) and (7.27) that $\lambda_1 = -q_{(0)}^2$ and $\omega_1 = 0$. Thus, the solutions of (7.26) and (7.27) are, respectively,

$$q_{(0)} = q_0 \operatorname{sech}[q_0(\tau + \overline{\beta}q_0^2\xi)]\exp(jq_0^2\xi/2) \tag{7.28}$$

$$q_{(1)} = -3jq_0^2 \operatorname{sech}[q_0(\tau + \overline{\beta}q_0^2\xi)]\tanh[q_0(\tau + \overline{\beta}q_0^2\xi)]\exp(jq_0^2\xi/2) \tag{7.29}$$

Hence, there is a quasi-stationary solution with the soliton part given by (7.28) and the nonsoliton part defined by (7.29) for the input pulse defined by the function $q(0, \tau) = q_0 \operatorname{sech}(q_0\tau)$. The third-order dispersion term causes modification of the soliton velocity by the factor $\overline{\beta}q_0^2$ and changes the soliton shape. But both these effects are of the order $\overline{\beta} \ll 1$, so we can say that the soliton is stable under the third-order dispersion term influence.

The influence of the loss on soliton behavior is stronger and can be analyzed by application of the perturbation method as well. Equation (7.4) with the included loss term takes the form

$$j\frac{\partial q}{\partial \xi} + \frac{1}{2}\frac{\partial^2 q}{\partial \tau^2} + |q|^2 q = -j\Gamma q \tag{7.30}$$

The perturbation term is just the loss term, $j\Gamma q$. By the expansion

$$q = q_{(0)} + \Gamma q_{(1)} + \ldots \tag{7.31}$$

and by the replacement $T = \Gamma\xi$, the following equation is obtained [1]:

$$j\frac{\partial q_{(1)}}{\partial \xi} + \frac{1}{2}\frac{\partial^2 q_{(1)}}{\partial \tau^2} + 2|q_{(0)}|^2 q_{(1)} + q_{(0)}^2 q_{(1)}^* = -j\frac{\partial q_{(0)}}{\partial T} - jq_{(0)} \tag{7.32}$$

The quasi-stationary solution of this equation is

$$q(\tau, \xi) = \eta \operatorname{sech}(\eta\tau)\exp(j\sigma) + j\frac{\eta\Gamma\tau^2}{2}\operatorname{sech}(\eta\tau)\exp(j\sigma) + \theta(\Gamma^2) \tag{7.33}$$

where

$$\eta = q_0\exp(-2\Gamma\xi) \tag{7.34}$$

and

$$\sigma = \frac{q_0^2}{8\Gamma}[1 - \exp(-4\Gamma\xi)] \tag{7.35}$$

Notice that the soliton amplitude decreases as $\exp(-2\Gamma\xi)$. At the same time, the total soliton pulse width increases and becomes

$$2t(\xi) = 2t_0\exp(2\Gamma\xi) \tag{7.36}$$

For a minimal optical loss in a single-mode optical fiber (0.2 dB/km), the soliton pulse doubles its width after about 15 km. The peak of optical power is proportional with

$$q^2 \simeq \exp(-4\Gamma\xi) \tag{7.37}$$

but the total energy is equal to

$$W = \int |q|^2 d\tau \tag{7.38}$$

Hence, the number of photons in a pulse decays proportionally with $\exp(-2\Gamma\xi)$. Since the number of photons is only important for photodetectors, the loss influence on the information signal level is the same as in a linear transmission system. We will take this fact into account during the BER calculation in a nonlinear system.

7.2.2 Influence of Inhomogeneity on Soliton Generation

It has been already mentioned that transmission of solitons in optical fibers must be in the anomalous dispersion wavelength region to achieve mutual compensation between dispersion and nonlinear effects. It is well known (see Figure 7.1) that the anomalous dispersion region lies above the wavelength $\lambda_m = 1.27$ μm for pure silica optical fibers. By proper design of a single-mode optical fiber, the point λ_m of zero (or, rather, minimal) dispersion can be moved toward the wavelength region of fiber-loss minimum. Such a fiber is called *dispersion-compensated fiber*. Since the dispersion minimum corresponds to the point $\lambda_m = 1.5$ μm in the dispersion-compensated fiber, we can say that the anomalous dispersion region lies above 1.5-μm wavelength.

The authors of [4] have shown that the soliton regime can exist even in the normal dispersion region for $\beta'' > 0$, as well. In such a case, the generated soliton envelope has the form

$$q(\tau, \xi) = aq_0[1 - a_1^2 \operatorname{sech}^2(q_0\tau)]^{1/2}, a_1 \le 1 \tag{7.39}$$

This envelope takes the form of a hyperbolic tangent function for $a_1 = 1$, as shown in Figure 7.3 for $a = q_0 = 1$.

It can be seen from Figure 7.3 that the hyperbolic tangent is characterized by the intensity gap in the waveform, so the solitons described by the hyperbolic tangent

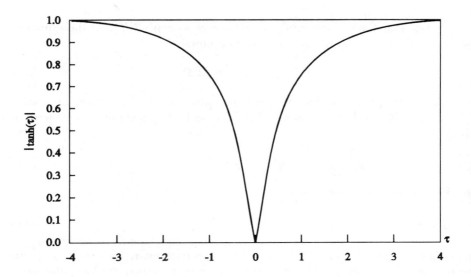

Figure 7.3 The function of hyperbolic tangent corresponding to the dark soliton.

function are called "dark solitons." It is clear that dark solitons have mainly theoretical meaning, as they cannot be used for carrying information in an optical fiber. For that reason, we will not analyze the characteristics of dark solitons, but it is interesting to discuss the possibilities for spreading the region of the bright-soliton existence.

It has been shown [5–7] that the bright-soliton region can be spread by influence of inhomogeneities in the optical fiber refractive-index profile. To describe this effect, the refractive-index profile in the optical fiber should be expressed as

$$n(r, \omega, E) = n_1(r, \omega) + n_2|E|^2 \tag{7.40}$$

where r is the radial coordinate. Hence, the radial variation of the refractive-index profile is taken into account, as well. It is useful to approximate function $n_1(r, \omega)$ by the product

$$n_1(r, \omega) = n_1(\omega)f(r) \tag{7.41}$$

The base form of the nonlinear differential equation that describes the soliton regime in inhomogeneous optical fibers can be obtained by applying the general form of the wave equation [5]

$$\nabla^2 E - \frac{1}{c^2} \frac{\partial^2 D}{\partial t^2} = 0 \tag{7.42}$$

where $\mathbf{E}$ and $\mathbf{D}$ are vectors of the electrical field and electrical induction, respectively. By inclusion of the nonlinear term, (7.42) becomes

$$\nabla^2 \mathbf{E} - \frac{1}{c^2} \frac{\partial^2 \mathbf{D}}{\partial t^2} = \frac{2n_0 n_2}{c^2} \frac{\partial^2}{\partial t^2} (|\mathbf{E}|^2 \mathbf{E}) \tag{7.43}$$

where $n_0 = n_1(0, \omega_0)$.

The following relationship between $\mathbf{D}$ and $\mathbf{E}$ is valid in an inhomogeneous medium [5]:

$$\mathbf{D} = f(r) \int n_1^2(t - t') \mathbf{E}(r, z, t') dt' \tag{7.44}$$

The electrical field vector can be expressed as a product of the radial and axial components:

$$\mathbf{E} = eu(z, t)\psi(r)\exp[j(\beta z - \omega t)] \tag{7.45}$$

so we have

$$-\frac{1}{c^2} \frac{\partial^2 \mathbf{D}}{\partial t^2} = ef(r)\exp[j(\beta z - \omega_0 t)]\int k^2(\omega + \omega_0)u(z, t)\psi(r)\exp(-j\omega t)d\omega \tag{7.46}$$

where

$$k^2(\omega + \omega_0) = \frac{(\omega + \omega_0)^2}{c^2} n_1^2(\omega + \omega_0) \tag{7.47}$$

The wavenumber k can be expanded to the power series in the vicinity of point $k(\omega_0) = k_0$. Applying only the two first terms of the expansion to (7.46), it becomes

$$-\frac{1}{c^2} \frac{\partial^2 \mathbf{D}}{\partial t^2} = ef(r)\exp[j(\beta z - \omega_0 t)]\left(k_0^2 + 2jk_0 k_0'' \frac{\partial}{\partial t} - b\frac{\partial^2}{\partial t^2}\right)u(z, t)\psi(r) \tag{7.48}$$

Equations (7.43), (7.45), and (7.48) are relevant for consideration of the soliton regime in an inhomogeneous optical fiber. From these equations, an approximate scalar wave equation can be obtained [6, 7], that is,

$$Q\Big(u(z, t)\psi(r)\Big) = 0 \tag{7.49}$$

where Q is an operator:

$$Q \equiv \frac{\partial^2}{\partial r^2} + \frac{\partial^2}{\partial z^2} - \beta^2 + 2j\beta \frac{\partial}{\partial z} + f(r)k_0^2 + 2ja(r)\frac{\partial}{\partial t}$$
$$- bf(r)\frac{\partial}{\partial t^2} + H|\psi(r)u(z, t)|^2 \tag{7.50}$$

where $b = k_0' + k_0 k_0''$, and $a = k_0 k_0' f(r)$.

Keeping in mind that the refractive index is only a function of the radial coordinate and that radial variations are independent on azimuthal ones, we can assume that the function $\psi(r)$ is a real one. The maximal value of $\psi(r)$ can be normalized to be equal unity. Such an assumption corresponds to the linear regime [8], but it is necessary for the elimination of the function $\psi(r)$ from (7.49). Equation (7.49) must be averaged over coordinate r to eliminate the function $\psi(r)$. After that, the equation equivalent to (7.1) is obtained. Since we have that

$$\int \psi(r)Q\left[u(z, t)\psi(r) \right]dr = 0 \tag{7.51}$$

(7.49) becomes

$$\left[d_1 - \beta^2 + \bar{f}k_0^2 + \frac{\partial^2}{\partial z^2} - b\bar{f}\frac{\partial^2}{\partial t^2} + d_2|u|^2 + 2jR \right]u(z, t) = 0 \tag{7.52}$$

where

$$R = \left(\beta\frac{\partial}{\partial z} + \bar{a}\frac{\partial}{\partial t} \right) \tag{7.53}$$

$$\bar{f} = \int \psi(r)f(r)\psi(r)dr \tag{7.54}$$

$$d_1 = \int \psi(r)\frac{\partial^2\psi(r)}{\partial r^2}dr \tag{7.55}$$

$$d_2 = \int H\psi^4(r)dr \tag{7.56}$$

Equation (7.52) is equivalent to (7.7) for an inhomogeneous optical fiber. The influence of the waveguide properties of the optical fiber is now included through the coefficients d_1 and d_2 and the average value of the refractive index function, $\bar{f}$.

To obtain the particular solution of (7.52) that eliminates self-phase modulation, the function $u(z, t)$ must be chosen to be the real one. At the same time it must be

$$\left(\beta\frac{\partial}{\partial z} + \bar{a}\frac{\partial}{\partial t}\right)u(z, t) = 0 \tag{7.57}$$

So we have

$$u = u(\xi), \ \xi = t - z/v_g \tag{7.58}$$

The pulse group velocity, v_g, now stands for

$$v_g = \beta/\bar{a} \tag{7.59}$$

The real part of (7.58) takes the form

$$\left[\left(\frac{\bar{a}^2}{\beta^2} - b\bar{f}\right)\frac{\partial^2}{\partial\xi^2} + d_2 u^2 + d_1 - \beta^2 + \bar{f}k_0^2\right]u(\xi) = 0 \tag{7.60}$$

This equation is a nonlinear, time-independent Schroedinger equation that gives the bright-soliton solution

$$u = u_0 \operatorname{sech}(\tau\xi) \tag{7.61}$$

if the following conditions are satisfied:

$$\beta^2 = k_0^2\bar{f} + d_1 + \frac{d_2 u_0^2}{2} \tag{7.62}$$

and

$$\left(\frac{\bar{a}^2}{\beta^2} - b\bar{f}\right)\tau^2 = \frac{d_2 u_0^2}{2} \tag{7.63}$$

Equation (7.60) also admits dark solitons, or

$$u = u_0[1 - a_1^2 \operatorname{sech}^2(\tau\xi)]^{1/2}, \ a_1 \le 1 \tag{7.64}$$

if the following conditions are satisfied:

$$\beta^2 = k_0^2\bar{f} + d_1 + d_2 u_0^2 \tag{7.65}$$

and

$$-\left(\frac{\overline{a}^2}{\beta^2} - b\overline{f}\right)\tau^2 = \frac{d_2 u_0^2}{2} \tag{7.66}$$

We can compare the conditions necessary for bright-soliton generation in homogeneous and in inhomogeneous optical fibers. The main difference is caused by the factor d_1 in (7.62) and (7.65), which determines the propagation constant. This factor actually causes the spreading of the bright-soliton existence area in an inhomogeneous optical fiber.

Since it must be that $\tau > 0$, the conditions for bright solitons given by (7.62) and (7.63) imply that

$$\frac{\overline{a}^2}{\beta^2 b \overline{f}} = \frac{k_0^2 \overline{f}}{(d_1 + 0.5 d_2 u_0^2)} \frac{1}{1 + R} > 1 \tag{7.67}$$

where

$$R = \frac{k_0 k_0''}{k_0'^2} \tag{7.68}$$

Because of the decreasing character of function $f(r)$, the factor d_1 is negative, so (7.68) will be satisfied only if $k_0'' > 0$. This condition points out the main difference between homogeneous and inhomogeneous optical fibers. In homogeneous optical fibers (when $d_1 = 0$), the bright solitons can exist only if $k_0'' < 0$.

The condition for dark-soliton generation in inhomogeneous optical fibers is

$$\frac{\overline{a}^2}{\beta^2 b \overline{f}} = \frac{k_0^2 \overline{f}}{k_0^2 \overline{f} + (d_1 + 0.5 d_2 u_0^2)} \frac{1}{1 + R} < 1 \tag{7.69}$$

and it can be satisfied both for $k_0'' > 0$ and for $k_0'' < 0$. After all, we can conclude that the zero dispersion condition is not $k_0'' = 0$ any more. The point of zero dispersion is moved toward lower wavelengths (below 1.27 μm). Since $d_1 < 0$, the zero dispersion condition can be achieved only if $k_0'' > 0$.

The previous conclusions can be illustrated by an example of parabolic inhomogeneity in a single-mode optical fiber. The parabolic refractive-index profile can be expressed as

$$f(r) = \begin{cases} 1 - r^2/L^2, & \text{for } r < r_0 \\ 1 - r_0^2/L^2, & \text{for } r \geq r_0 \end{cases} \tag{7.70}$$

where L is the length dimensional constant, and r_0 corresponds to the core radius of the single-mode optical fiber. The wave function, $\psi(r)$, for this profile can be approximated by

$$\psi(r) = \exp\left[-\frac{k_0 r^2}{2L}\right] \tag{7.71}$$

The bright-soliton solution in such a fiber can exist if the following conditions are satisfied:

$$\beta^2 = k_0^2\left(\bar{f} - \frac{1}{k_0 L}\right) + \frac{Hu_0^2}{4} \tag{7.72}$$

$$\tau^2\left(\frac{\bar{a}^2}{\beta^2} - b\bar{f}\right) = \frac{Hu_0^2}{4} \tag{7.73}$$

and

$$\bar{f} = \frac{1}{k_0 L}\left[\exp\left(-\frac{k_0 r_0^2}{L}\right) - 1\right] + 1 \tag{7.74}$$

The zero-dispersion condition for the refractive-index profile (7.70) becomes

$$R = \frac{k_0 k_0''}{k_0'^2} \simeq \frac{1}{k_0 L \bar{f}} \tag{7.75}$$

Since $k_0 r_0^2/L \gg 1$, (7.75) can be approximated by

$$R \simeq \frac{1}{k_0 L - 1} \tag{7.76}$$

For wavelength $\lambda = 1$ μm, it will be $k_0 L \simeq 0.02$. Assuming that $(r_0/L)^2 = 2\%$, the core radius becomes $r_0 \simeq 4.5$ μm. The approximation in (7.76) is not valid for large values of the ratio (r_0/L). The functional dependence (7.75)—or, rather, (7.76)—can be expressed as the function $L = f(\lambda)$. This function is shown in Figure 7.4 for dispersion-compensated single-mode optical fibers. The function $L = f(\lambda)$ separates the regions of bright-end dark-soliton existence. Every couple (L, λ) above the curve in Figure 7.4 causes dark-soliton appearance, while every couple (L, λ) below the curve causes bright-soliton appearance.

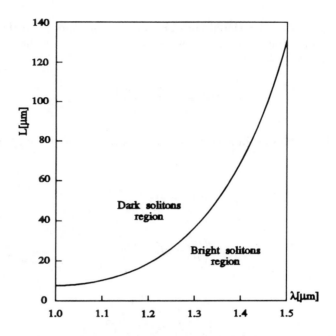

Figure 7.4 The bright and dark soliton regions in the dispersion-compensated single-mode optical fiber with parabolic refractive-index profile.

7.2.3 Mutual Interaction Among Solitons in Optical Fibers

So far, only a separate soliton pulse behavior in the optical fiber has been discussed. But it is well known that transmission of information is always made by a pulse train. Thus, it is important to investigate what will happen in the soliton pulse train during its propagation through an optical fiber. Practically speaking, that means considering the mutual influence among neighboring soliton pulses. It has been shown that this mutual influence among neighboring solitons cannot be neglected and must be taken into account [2]. A graphical illustration of the mutual interaction between two solitons is given in Figure 7.5.

To prevent such interaction, the soliton separation must be maintained to be greater than some critical value. The eventual mutual interaction among solitons will cause their coalescence and information loss [9–11]. The exact analysis of the mutual interaction between solitons is complex, and its presentation is aimless. It is more useful to consider the propagation of two soliton pulses being the solutions of (7.7) and examine the factors that influence the mutual interaction [10–12]. Therefore, it is important to search for methods to maintain soliton separation to prevent their interaction.

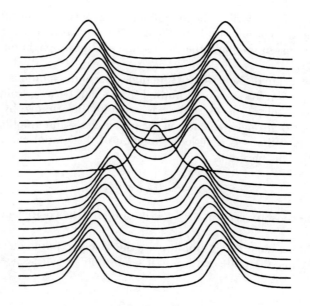

Figure 7.5 The illustration of mutual interactions among solitons in the optical fiber.

Several methods have been suggested to minimize the mutual interaction among solitons [10–13], which we will present and comment on their main results. The solution of (7.7) for two input pulses can be obtained by employment of the inverse-scattering method, as in [2] and [11], that is,

$$q(\tau, \xi) = Q\bigg\{ \eta_1 \operatorname{sech}[\eta_1(\tau + \Delta\tau)]\exp(j\eta_1^2\xi)$$
$$+ \eta_2 \operatorname{sech}[\eta_2(\tau - \Delta\tau)]\exp(j\eta_1^2\xi)\bigg\} \tag{7.77}$$

where

$$Q = \left\{ \frac{(\eta_1^2 - \eta_2^2)}{(\eta_1^2 + \eta_2^2) - 2[(\eta_1^2 - P_1^2)^{1/2}(\eta_2^2 - P_2^2)^{1/2} + P_{1,2}\cos\psi]} \right\} \tag{7.78}$$

$$P_{1,2} = \eta_{1,2}\operatorname{sech}[\eta_{1,2}(\tau \pm \Delta\tau)] \tag{7.79}$$

$$\psi = \frac{(\eta_1^2 - \eta_2^2)\xi}{2} \tag{7.80}$$

The function in (7.77) describes the interactions among two solitons for any distance between the solitons. The amplitude coefficient, Q, is a measure of the mutual interaction, while the terms in the parentheses describe soliton propagation in the absence of interaction.

Thus, the exact hyperbolic secant functions are not the solutions of the nonlinear Schroedinger equation in the case of mutual interaction between solitons. The solutions are the quasi-soliton pulses with shapes mainly determined by the coefficient Q. The coefficient Q strongly depends on the parameter τ and amplitudes η_1 and η_2. The separation of solitons is a periodic function of the spatial variable ξ [through $\cos \psi$ in (7.78)]. This period is

$$\Lambda = 4\pi/(\eta_1^2 - \eta_2^2) \tag{7.81}$$

To maintain the constant separation between solitons, the coefficient $\cos \psi$ must be reduced.

When solitons have equal amplitudes, the distance parameter, $\Delta\tau$, decreases with the time. Thus, the pulses with initial separation τ_0 coalescence into one pulse after a period

$$T = \frac{\pi \cosh^2(\tau_0)}{[3 + \cosh^2(\tau_0)]^{1/2}} \tag{7.82}$$

In such a case, the mutual interaction among pulses cannot be prevented. But the initial separation can be large enough to prevent the interaction over the prescribed length. It is clear that this length corresponds to the regeneration distance, which will be discussed in Chapter 9.

When the solitons have unequal amplitudes, the distance parameter, $\Delta\tau$, is given by

$$\Delta\tau \simeq \tau_0 - \frac{1+k}{2}\ln\left(\frac{1+k}{1-k}\right) \tag{7.83}$$

In this case, two solitons basically maintain their initial separation. The separation between solitons is more stable for a larger difference in the soliton amplitude. Thus, it might be an effective way to maintain constant the separation between neighboring pulses in a soliton train. But we must remember that different pulse amplitudes induce different group velocities of the pulses, which will degrade the positive effect obtained. That fact will be discussed in Chapter 9.

Another way to prevent the mutual interaction among solitons is by the influence of third-order dispersion term. Numerical calculations of (7.4) (but without loss term, as in [12]) show that third-order dispersion term reduces the mutual interaction among bright solitons. The normalized dispersion parameter, $\overline{\beta}$, defined as

$$\overline{\beta} = \frac{1}{6} \frac{\beta'''}{\beta'' t_0} \tag{7.84}$$

is a measure of dispersion influence on the mutual interaction among solitons [see (7.3b) and (7.6)]. The value t_0 in (7.84) refers to the pulse half-width. It was shown that the value $\overline{\beta} = 0.05$ practically prevents the mutual interaction among solitons [12]. Since $\overline{\beta}$ depends on the optical fiber parameters, the proper design of the optical fiber operating regime can prevent the mutual interaction among solitons. Most important is the choice of the operating wavelength, while a small correction is possible by the initial pulsewidth. For example, for silica-core optical fiber with zero-dispersion wavelength $\lambda_0 = 1.27$ μm, operation at $\lambda = 1.271$ μm gives $\beta'' = 0.09$ ps^2/km and $\beta''' = 0.054$ ps^3/km. To achieve the value $\overline{\beta} = 0.05$, the initial pulsewidth should be $2t_0 = 4$ ps. But if the wavelength $\lambda = 1.278$ μm is chosen, the initial pulsewidth should be 2 ps.

7.3 USE OF GAUSSIAN PULSES INSTEAD OF SOLITONS

The mutual interactions among solitons in the pulse train drastically decrease the transmission capacity of the optical fiber [13]. That is caused by the fact that an initial separation between the solitons must be kept much larger than the separation initiated by the transmission rate. That was the reason for the proposal to use a Gaussian pulse [14],

$$q(\tau, \xi) = cq_0 \exp[-(q_0\tau/\mu)^2]\exp(jq_0^2\xi/2) \tag{7.85}$$

instead of a soliton pulse,

$$q(\tau, \xi) = q_0 \, \text{sech}(q_0\tau)\exp(jq_0^2\xi/2) \tag{7.86}$$

The Gaussian pulse with the shape given by (7.85) is the approximate solution of (7.7). The parameters c and μ in (7.85) are chosen in such a way that the Gaussian pulse has the energy equal to the soliton energy. Hence, the areas under the corresponding curves must be equal, as shown in Figure 7.6.

Numerical solutions of (7.7) with a Gaussian function as an input show that the Gaussian pulses also propagate through the optical fiber with a small change of shape. The amplitude of a Gaussian pulse is accompanied by an oscillatory tail, the amplitude of which increases with the incident pulse amplitude.

During the propagation of the Gaussian pulses, the mutual interaction between neighboring pulses is observed as having the nature similar to the soliton interaction. But the period of coalescence is at least 50% larger than the same period in the soliton interaction. On the other hand, this period increases exponentially with the

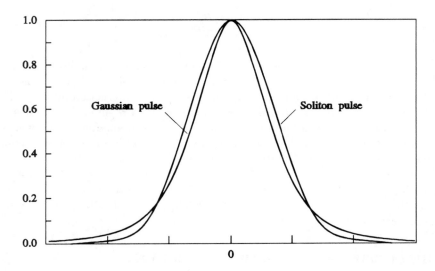

Figure 7.6 The shape of pulses with equivalent energy.

increase of initial separation between the neighboring pulses. This implies that the transmission of Gaussian pulses provides at least 50% more transmission rate than solitons with the same energy. Another advantage is related to pulse generation. It is well known that the generation of Gaussian pulses is simpler than generation of solitons [15]. So, in conclusion, we can say that it is convenient to use Gaussian pulses instead of hyperbolic-secant-type solitons for signal transmission in nonlinear lightwave systems.

The main theoretical results related to soliton generation were experimentally proved in 1980 [16]. The possibility of long-length soliton transmission was experimentally verified a little later [17–19].

7.4 SOLITON SELF-FREQUENCY SHIFT EFFECT

The Raman effect causes continuous downshift of the carrier optical frequency of the soliton pulses propagating in optical fiber. This effect has been described theoretically by Gordon [20], who found that the shift strongly depends on the pulsewidth, t_0, and scales linearly with soliton optical power. According to the Gordon theory, the optical frequency, f, of the pulse propagating in fused silica optical fiber can be expressed as

$$\frac{df}{dz} = \frac{0.0436}{t_0^{\,4}} \lambda'^2 D' P' \qquad (7.87)$$

where the pulsewidth, t_0, is expressed in picoseconds, normalized wavelength $\lambda' = \lambda/1.5$ in micrometers and normalized dispersion $D' = D/15$ in picoseconds per nanometer kilometer, while P' represents the relative pulse power in relation to the peak power value.

The self-frequency shift changes the timing of the pulses in a nonlinear lightwave system leading to time jitter. Besides, this shift couples amplitude fluctuations to frequency and timing fluctuations. Because of that, self-frequency shift is one of the factors limiting transmission capacity (expressed through length bit rate product) of the nonlinear lightwave system. If there is a chain of in-line optical amplifiers (especially erbium-doped fiber amplifiers; see Chapter 8), the nonlinear system capacity is also limited by the spontaneous emission that always accompanies coherent amplification. As we can see from Section 9.2.2, when the noise field is added to the optical soliton field, a part of the noise is incorporated into the soliton. One of the resulting effects is a random shift of the soliton carrier frequency, with a corresponding change in the soliton pulse velocity. This effect has been studied by Gordon and Haus [21], so the effect is known as the *Gordon-Haus effect*. This random walk can cause timing errors as well and limit the total transmission capacity of the nonlinear lightwave system by limiting the spacing between neighboring in-line optical amplifiers.

The random shift in the soliton carrier frequency is calculated in [21] and can be expressed as

$$\Delta\omega = 2\pi\,\Delta f = \frac{2}{3}a\eta \tag{7.88}$$

where a reflects the nonsoliton (noise) part of the pulse amplitude, while η represents the soliton part of the pulse amplitude. The authors of [21] evaluated the time shift of arrival times of soliton pulses in a pulse train. The variance of the time arrival times can be expressed as

$$\sigma^2 \cong \frac{\gamma\eta L^3}{9N} \tag{7.89}$$

where γ is the optical power loss per unit length, η is the soliton pulse amplitude, L is the total transmission line distance, and N is the number of photons per unit of energy.

To increase the transmission capacity of a nonlinear lightwave system, the Gordon-Haus effect should be suppressed. That can be done by using narrowband frequency-guiding filters periodically distributed along the transmission line. It was shown in [22] that if the solitons are continuously passed through narrowband optical filters, the frequency fluctuations reach a steady-state level. In such a way, the Gordon-Haus limit is partially suppressed, and system capacity will be increased. If optical filters are inserted along the transmission line, some additional loss on the transmission line will be observed. Hence, an extra amplifier must be employed to offset the loss experienced, which will cause spontaneous emission noise to rise exponentially with the distance (see Section 8.4). This all will limit the benefit in transmission system capacity brought by the insertion of optical filters.

Soliton jitter can be suppressed by use of a sliding frequency-guiding filter [23], where the peak frequency of the filters is gradually translated along the optical transmission line. In such a way, the optical transmission line becomes transparent for the optical solitons while being the efficient noise filter. This method allows employment of stronger filters without incurring the penalty of exponentially rising spontaneous emission noise, as was the case in the aforementioned method. Both stationary-filter and sliding-frequency-filter suppression methods are applicable to a picosecond pulse transmission nonlinear system, while in a subpicosecond pulse system, a broad pulse bandwidth restricts the strength of the filters that can be employed.

A third method has been proposed for the suppression of Gordon-Haus time jitter. This jitter can be partially overcome by the combination of in-line filtering and receiver dispersion compensation [24]. It has been shown that even if the receiver dispersion compensation method becomes less efficient with narrow filters broadening, the optimum choice of system parameters allows an effective reduction of Gordon-Haus jitter by a factor of 3.5.

REFERENCES

[1] Hasegawa, A., and Y. Kodama, "Signal Transmission by Optical Solitons in Monomode Fiber," *IRE Proc.*, 69(1981), pp. 1145–1150.

[2] Bullough, R., and P. J. Caudrey, *Solitons*, Berlin: Springer-Verlag, 1980.

[3] Kodama, Y., "Higher Order Approximation in the Reductive Perturbation Method. II. The Strongly Dispersive System," *J. Phys. Soc. Jap.*, 45(1978), pp. 311–315.

[4] Hasegawa, A., and F. Tapert, "Transmission of Stationary Nonlinear Optical Pulses in Dispersive Dielectric Fibers. II. Normal Dispersion, *Appl. Phys. Lett.*, 15(1973), pp. 171–172.

[5] Jain, M., and N. Tzoar, "Propagation of Nonlinear Optical Pulses in Inhomogeneous Media," *J. Appl. Phys.*, 49(1978), pp. 4649–4653.

[6] Bendow, B., et al., "Theory of Nonlinear Pulse Propagation in Optical Waveguides," *J. Opt. Soc. Amer.*, 70(1980), pp. 539–546.

[7] Jain, M., and N. Tzoar, "Nonlinear Pulse Propagation in Optical Fibers," *Optics Lett.*, 3(1978), pp. 202–204.

[8] Marcuse, D., and C. Lin, "Low Dispersion Single-Mode Fiber Transmission. The Question of Practical Versus Theoretical Maximum Transmission Bandwidth," *IEEE J. Quantum Electron.*, QE-17(1981), pp. 1012–1017.

[9] Doran, N., and K. Blow, "Solitons in Optical Communications," *IEEE J. Quantum Electron.*, QE-19(1983), pp. 1883–1888.

[10] Smith, K., and L. Molenauer, "Experimental Observation of Soliton Interaction Over Long Fiber Paths: Discovery of Long Range Interaction," *Optics Lett.*, 14(1989), pp. 1284–1286.

[11] Chu, P. L., and C. Desem, "Mutual Interaction Between Solitons of Unequal Amplitudes in Optical Fibre," *Electron. Lett.*, 21(1985), pp. 1133–1134.

[12] Chu, P. L., and C. Desem, "Effect of Third Order Dispersion of Optical Fibre on Soliton Interaction," *Electron. Lett.*, 21(1985), pp. 228–229.

[13] Shiojiri, E., and Y. Fuji, "Transmission Capability of an Optical Fiber Communication System Using Index Nonlinearity," *Appl. Opt.*, 24(1985), pp. 358–360.

[14] Chu, P. L., and C. Desem, "Gaussian Pulse Propagation in Nonlinear Optical Fiber," *Electron. Lett.*, 19 (1983), pp. 956–957.

[15] Hong, B. J., et al., "Using Nonsoliton Pulses for Soliton Based Communications," *IEEE/OSA J. Lightwave Techn.*, LT-8(1990), pp. 568–575.

[16] Mollenauer, L. F., et al., "Experimental Observation of Picosecond Pulse Narrowing and Solitons in Optical Fibers," *Phys. Rev. Lett.*, 45(1980), pp. 1095–1096.

[17] Beaud, P., et al., "Ultrashort Pulse Propagation, Pulse Breakup, and Fundamental Soliton Formation in a Single Mode Optical Fiber," *IEEE J. Quantum Electron.*, QE-23(1987), pp. 1938–1946.

[18] Mollenauer, L. F., and K. Smith, "Demonstration of Soliton Transmission Over More Than 4000 km in Fiber With Loss Periodically Compensated by Raman Gain," *Optics Lett.*, 13(1988), pp. 675–677.

[19] Taga, H., et al., "5 Gb/s Optical Soliton Transmission Experiment Over 3000 km Employing 91 Cascaded Er-Doped Fibre Amplifier Repeaters," *Proc., ECOC'92*, Berlin, 1992, pp. 855–858.

[20] Gordon, J. P., "Theory of the Soliton Self-Frequency Shift," *Optics Lett.*, 11(1986), pp. 662–664.

[21] Gordon, J. P., and H. A. Haus, "Random Walk of Coherently Amplified Solitons in Optical Fiber Transmission," *Optics Lett.*, 11(1986), pp. 665–667.

[22] Kodama, Y., and A. Hasegawa, "Generation of Asymptotically Stable Optical Solitons and Suppression of the Gordon Haus Effect," *Optics Lett.*, 17(1992), pp. 31–33.

[23] Molenauer, L. F., et al., "The Sliding Frequency Guiding Filter: An Improved Form of Soliton Jitter Control," *Optics Lett.*, 17(1992), pp. 1575–1577.

[24] Chandrakumar, V., et al., "Combination of In-Line Filtering and Receiver Dispersion Compensation for Optimized Soliton Transmission," *IEEE/OSA J. Lightwave Techn.*, LT-12(1994), pp. 1047–1051.

Chapter 8

Optical Amplifiers for Coherent and Nonlinear Systems

8.1 INTRODUCTION

Optical amplifiers are the devices where the direct amplification of incident optical signals is performed. That means that the amplification is performed on the optical level, without any optoelectronic conversion. Optical amplifiers can be used in both coherent and nonlinear lightwave systems (i.e., DD systems). In a lightwave telecommunications system, the optical amplifier can be used at three separate places. The first place is in the optical transmitter to compensate for the losses due to coupling or modulation and to raise the transmitted level of the optical power. This application refers primarily to coherent optical transmitters. The second place is on an optical line to compensate for the optical fiber loss; such an application refers to both coherent and nonlinear lightwave systems. The in-line optical amplifiers in a coherent optical system perform a periodic increase in the optical power level. Their application in a single-channel nonlinear lightwave system means not only amplification but full regeneration of the pulses, as well. Actually, both the amplification and the regeneration happen through the reshaping of optical pulses. The third place an optical amplifier can be used is in the optical receiver as a preamplifier stage. This application means that the optical signal is first amplified by the amplifier and then detected by a photodiode. In such a way, the preamplifier reduces the effects of receiver thermal noise and improves receiver sensitivity.

All optical amplifiers can be classified into three main groups: semiconductor laser amplifiers, amplifiers based on stimulated scattering effect, and doped optical-fiber amplifiers. A laser amplifier's operation is based on the effect of the stimulated emission of light signals under the influence of the current pump. The emitted light signal has the same frequency and phase as an initial attenuated optical signal that reaches the laser structure. The laser amplifier is pumped by a direct-current signal having a value near the lasing threshold. The amplification of the light signal, based on the stimulated emission, has a linear character, so these amplifiers are often called linear amplifiers. A subgroup of laser amplifiers are injection-locked laser amplifiers. (Their operation is explained in Chapter 3.) Such amplifiers can be used only for amplification of frequency-modulated or phase-modulated optical signals. The pumping of injection-locked amplifiers is made by current under the threshold value.

Linear laser amplifiers can be classified into Fabry-Perot–type amplifiers and traveling-wave amplifiers, depending on the structure of the optical resonator (or optical cavity). The difference between these types is related mainly to the reflectivity of cavity facets. The facet reflectivity of Fabry-Perot amplifiers is large; thus, the optical feedback in the cavity is significant. The stimulated emission is induced by an incident optical signal coming into the cavity. Under the influence of bias current (a little above the threshold) and the incident signal, the semiconductor structure falls into the lasing regime, and a classical laser emission occurs. Thus, the amplification of the incident optical signal is a result of multiway traveling between the facets in the resonator. The maximum signal gain in such a structure is limited by the saturation regime, when the amplified signal level reaches the saturation output power from the semiconductor structure [1, 2]. The noise in Fabry-Perot laser amplifiers appears due to spontaneous emission generated in the laser, a process that will be analyzed in detail.

Traveling-wave semiconductor amplifiers possess many advantages compared with Fabry-Perot amplifiers, such as broadband signal gain, high saturation output power, and small noise figure. The limitation due to the resonant nature of Fabry-Perot amplifiers is overcome by a reduction of optical feedback in the cavity, which is achieved by application of antireflection coatings to the resonator facets. The antireflection coating prevents any internal reflection in the cavity; thus, we can say that the amplification is a result of only one traveling of light through the resonant cavity. Traveling-wave amplifiers ensures more stable gain characteristics than Fabry-Perot amplifiers, but a relatively large bias current is necessary to achieve a gain comparable with the gain of Fabry-Perot amplifiers [3, 4].

Optical amplifiers based on the stimulated scattering effect can use both the forward and the backward scattered waves for amplification of the incident optical signal. (These scattering effects were explained in Chapter 6.) Such amplifiers are pumped by the optical signal having a wavelength lower than the wavelength of the incident optical signal that is to be amplified. It has been shown that Raman amplifiers with forward scattering are the most suitable for practical application, because of a relatively large amplification coefficient and wide bandwidth. The main imperfection of Brillouin-type amplifiers is a rather narrow bandwidth of gain. Both Raman-type and Brillouin-type amplifiers can be used in lightwave systems, although there are attempts to apply amplifiers based on four-photon mixing for optical signal amplification [5–7], as well. The noise appearing in optical amplifiers based on stimulated emission is actually the signal due to spontaneous emission, which will be considered in detail.

The third and, seemingly, most important group of optical amplifiers is doped optical-fiber amplifiers. Although they can be considered a special class of traveling-wave amplifiers, we will treat these amplifiers independently, because of their importance for both nonlinear and coherent lightwave systems, in Section 8.4. For now we will just say that doped optical amplifiers are a kind of optical-fiber lasers in which the appropriate outside element ions are implemented into an optical-fiber glass structure [8]. Optical-fiber amplifiers doped with rare-earth-element ions are now recognized as essential and practical components for lightwave communications systems operating in the 1550-wavelength region. Erbium-doped optical-fiber amplifiers play a special role because of their favorable characteristics, which have been verified in a number of experimental realizations.

The main parameters that characterize optical amplifiers are amplification coefficient, noise figure, and amplification bandwidth. There is a mutual correlation among these parameters. Based on common experience, it will be most useful to consider the amplifier noise characteristics first and then to make relevant conclusions.

8.2 NOISE CHARACTERISTICS OF OPTICAL AMPLIFIERS

8.2.1 Noise Characteristics of Laser Amplifiers

It is most suitable to consider the noise characteristics of laser amplifiers on a two-level atomic model (level 1 and level 2), by using probabilities of photon emission and photon absorption in unit time intervals [9, 10]. These probabilities can be denoted by α and β, respectively. According to the definition equations from Chapter 3, we have

$$\alpha = A_{12}N_2 \tag{8.1a}$$

$$\beta = A_{12}N_1 \tag{8.1b}$$

where A_{12} is the Einstein coefficient of spontaneous emission (or absorption), while N_1 and N_2 denote the number of electrons on the lower (1) level and the higher (2) level, respectively. Thus, the probability that the number of electrons will increase from the value $n - 1$ to value n by spontaneous and stimulated emissions is equal to αn. On the other hand, the probability that the electron number will decrease from the value $n + 1$ to the value n by stimulated absorption is $\beta(n + 1)$. In the same way, the probability of the number of electrons increasing from n to $n + 1$ is $\alpha(n + 1)$, while the probability of the electron number decreasing from n to $n - 1$ is βn. By taking the corresponding contributions, the differential equation that describes the probability, P_n, that a finding of n photons in an amplified light signal is

$$\frac{dP_n}{dt} = \beta(n + 1)P_{n+1} - \alpha(n + 1)P_n + \alpha n P_{n-1} - \beta n P_n \tag{8.2}$$

To find the noise power, it is necessary to define the mean value and the variance of the photon number n. The mean value $<n>$ and the variance $<n^2>$ of the photon number are defined by

$$<n> = \sum_n n P_n \tag{8.3}$$

$$<n^2> = \sum_n n^2 P_n \tag{8.4}$$

These values can be found if we know the probability, P_n. However, it is more useful to use (8.2) and establish the following equations for the mean value and the variance [9]:

$$\frac{d<n>}{dt} = (\alpha - \beta)<n> + \alpha \tag{8.5}$$

$$\frac{d<n^2>}{dt} = 2(\alpha - \beta)<n^2> + (3\alpha + \beta)<n> + \alpha \tag{8.6}$$

When (8.5) and (8.6) are solved for the initial conditions $<n(0)> = \overline{n}_0$ and $<n^2(0)> = \overline{(n_0)}^2$, the following solutions are obtained:

$$<n> = G\overline{n}_0 + (G - 1)n_{sp} \tag{8.7}$$

$$\begin{aligned} <n^2> = \sigma^2 &= G\overline{n}_0 + (G - 1)n_{sp} + (G - 1)^2 n_{sp}^2 \\ &+ 2\overline{n}_0 G(G - 1)n_{sp} + G^2(\overline{n_0^2} - \overline{n}_0^2 - \overline{n}_0) \end{aligned} \tag{8.8}$$

where G and n_{sp} are given by

$$G = \exp[(\alpha - \beta)t] \tag{8.9}$$

and

$$n_{sp} = \frac{\alpha}{\alpha - \beta} \tag{8.10}$$

The parameters G and n_{sp} represent the gain and the spontaneous emission factors, respectively.

The results of (8.7) and (8.8) are related to the single-wavelength optical signal. That wavelength is defined by the energy gap between levels 1 and 2. The mean value $<N>$ and the variance Σ^2 of the total number of photons in a real amplified optical signal can be determined by integration over all the frequency (or wavelength) region, so we have

$$<N> = \int_0^\infty <n>df \tag{8.11}$$

and

$$\Sigma^2 = \int_0^\infty \sigma^2 df \tag{8.12}$$

Applying the integration on (8.7) and (8.8), it can be obtained that

$$<N> = G\overline{N}_0 + (G + 1)n_{sp}\Delta f \tag{8.13}$$

and

$$\begin{aligned} \Sigma^2 = G\overline{N}_0 &+ (G - 1)n_{sp}\Delta f + (G - 1)^2 n_{sp}^2 \Delta f \\ &+ 2\overline{N}_0 G(G - 1)n_{sp} + G^2(\overline{N_0^2} - \overline{N}_0^2 - \overline{N}_0) \end{aligned} \tag{8.14}$$

where

$$N_0 = \int_0^\infty \overline{n}_0 df \tag{8.15}$$

and

$$\overline{N_0^2} = \int_0^\infty \overline{n_0^2} df \tag{8.16}$$

are related to the mean value and the variance of the photons in an incident optical signal. At the same time, the frequency region Δf is related to the width of the spectral line of the optical signal at the amplifier output.

The variance Σ^2 is a measure of the noise power that appears during the amplification process in the semiconductor structure. The first term on the right side of (8.14) represents the quantum noise caused by stimulated emission of the light, the second term is the quantum noise caused by spontaneous emission, the third term is the autocorrelational noise due to spontaneous emission, the fourth term is the cross-correlational noise due to the interaction of the signal and spontaneously emitted light, while the fifth term is the noise due to the noncoherent nature of the light. This last term is always relatively small and can be neglected.

In such a way, (8.13) and (8.14) can be used to evaluate the amplifier noise. It is most suitable to make an evaluation by the standard method at the optical-receiver end. The main parameter incorporating the amplifier-noise influence is the SNR at the optical-amplifier output or, rather, at the receiver photodiode output. The SNR, R_0, at the receiver photodiode output, for a nonamplified optical signal is given as (see Appendix A)

$$R_0 = \frac{[\overline{N}_0 q]^2}{2q^2 \overline{N}_0 B} = \frac{\overline{N}_0}{2B} \tag{8.17}$$

where B is the receiver filter bandwidth, and q is the electron charge. The SNR, R_1, for the amplified optical signal can be expressed as

$$R_1 = \frac{[<N>]^2}{2\Sigma^2 B} \tag{8.18}$$

To evaluate the ratio R_1, according to (8.18), we can use (8.13) and (8.14). Since the value of gain, G, is relatively large, and the mean number of photons is much higher than the spontaneous emission factor, the ratio R_1 can be expressed as

$$R_1 = \frac{\overline{N}_0^2}{2(2\overline{N}_0 n_{sp} + n_{sp}^2 \Delta f)B} \approx \frac{\overline{N}_0}{4 n_{sp} B} \tag{8.19}$$

The approximation in (8.19) is made under the assumption that the frequency range, Δf, is very narrow. Since the spontaneous emission factor, n_{sp}, has a value above unity for semiconductor lasers, it is clear that the SNR for the amplified signal is less than the SNR for a nonamplified signal. The deterioration can be measured by the ratio $r = 10\ell og[R_0/R_1]$, which has the limit value $r_1 = 10\ell og[2n_{sp}]$. If the spontaneous emission factor, n_{sp}, is just above unity, the ratio R_0/R_1, reflecting the amplifier noise figure, is 3 dB. The limit value of ratio R_0/R_1 refers to an ideal traveling-wave optical amplifier, but it can be assumed that the 3-dB value is the limit value for the most practical realizations of laser amplifiers [11].

8.2.2 Noise Characteristics of Raman Amplifiers

The noise characteristics of optical amplifiers based on stimulated scattering effects will be considered on the example of the Raman amplifier. It is clear that the spontaneously scattered light plays the role of noise in a Raman amplifier [12]. Since that light also will be amplified during its propagation, it should be taken into account.

The SNR can be characterized by the parameter R_r, defined as the ratio of the optical power of an amplified incident signal to the optical power of spontaneously scattered light at length z on the optical-fiber line, or

$$R_r = 10\ell\text{og}(r_r) = 10\ell\text{og}\left(\frac{P_s(z)}{P_{sp}(z)}\right) \tag{8.20}$$

where $P_s(z)$ and $P_{sp}(z)$ refer to the stimulated and the spontaneous signal power, respectively.

The ratio R_r for forward scattering will be found under the same assumptions as those in Chapter 6. Hence, the pump with power $P_p(0)$ is injected at the input point $z = 0$ of the optical-fiber line and propagates in the z^+ direction. The effective power density of the pump signal is $S_p(z) = P_p(z)/S$, where S is the cross-sectional area of the optical-fiber core. The total power of the spontaneously scattered signal can be determined by applying the expression for the total number of photons, $N_{sp}(L)$, in the Stokes mode due to spontaneous Raman scattering [see (6.13)]:

$$N_{sp}(L) = \exp(-\gamma L)\int_0^L \alpha S_p(0)\exp\left\{\frac{\alpha S_p(0)}{\gamma}[\exp(-\gamma\xi)-\exp(-\gamma L)]\right\}d\xi \tag{8.21}$$

where $\alpha = \alpha(f-f_p)$ is the Raman amplification coefficient, depending on the frequency, f; f_p is the frequency of the pump; and γ is the loss coefficient in the optical fiber. It is commonly assumed that the losses of the pump and the scattered signal in the optical fiber are the same, so the following relation is valid:

$$S_p(z) = S_p(0)\exp(-\gamma z) \tag{8.22}$$

The total power of the signal due to spontaneous Raman scattering can be found by integration of the number of photons over all longitudinal modes, that is,

$$P_{sp}(L) = \int_{f_p}^{\infty} hfN_{sp}(L)df \tag{8.23}$$

Where h is Planck's constant. Since the Raman amplification coefficient, α, has almost constant value in the vicinity of its maximum, we can assume that the Raman amplification coefficient is constant in the wavelength region that includes the amplified signal wavelength. This region can be defined by an optical filter employed after the amplifier. Hence, it is justified to assume that the Raman amplification coefficient has the constant value α_0 in the wavelength range $\Delta\lambda$ of interest. Under this assumption, (8.23) becomes

$$P_{sp}(L) = \frac{\Delta \lambda h c^2}{\lambda_0^3} \alpha_0 S_p(0) \exp(-\gamma L) \int_0^L \exp\left\{A[\exp(-\gamma \xi)-\exp(-\gamma L)]\right\}d\xi \quad (8.24)$$

where λ_0 is the wavelength corresponding to the maximum Raman amplification coefficient, and c is the light velocity. The integration over the wavelength range is performed under the assumption that $\lambda_0 \gg \Delta \lambda$. The parameter A is very important and expresses the net Raman amplification, that is,

$$A = \frac{\alpha_0 S_p(0)}{\gamma} \quad (8.25)$$

The largest contribution to the integral in (8.23) will occur while $\gamma \xi < 1$. This situation corresponds to the case when stimulated emission is predominant, or when $A \gg 1$. Next, the approximations $\gamma L \gg 1$ and $\exp(-\gamma \xi) = 1 - \gamma \xi$ are also justified for this case. Thus, (8.23) can be simplified and rewritten as

$$P_{sp}(L) = \Lambda \exp(-\gamma L + A) \quad (8.26a)$$

where

$$\Lambda = \frac{\Delta \lambda h c^2}{\lambda_0^3} \quad (8.26b)$$

is a constant, depending on the operating wavelength and the optical filter bandwidth.

At the same time, the power of useful stimulated scattered signal can be expressed as [see (6.5)]

$$P_s(L) = P_s(0) \exp(-\gamma L + A) \quad (8.27)$$

where $P_s(0)$ is the input signal power. Now the SNR in a Raman amplifier can be evaluated from (8.26) and (8.27), that is,

$$R_{rF} = 10\ell og(r_{rF}) = 10\ell og\left[\frac{P_s(L)}{P_{sp}(L)}\right] = 10\ell og\left[\frac{P_s(0)}{\Lambda}\right] \quad (8.28)$$

Some important conclusions can be made from (8.28). First, the SNR does not depend on the optical line length, but only on the input signal power. Second, the SNR increases with the wavelength of the scattered signal. The SNR is shown in Figure 8.1 for two values of the amplification wavelength region $\Delta \lambda$ (0.1 nm and 10 nm) and for scattered signal wavelengths 1.3 and 1.5 μm. The SNR has a value of about 30 to 40 dB for the input power of 1 mW (or 0 dB). The results from Figure 8.1

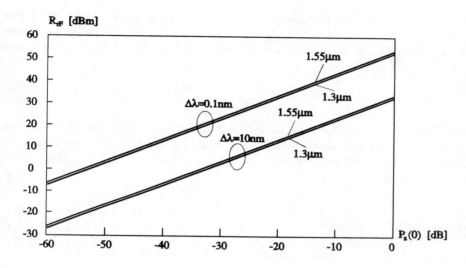

Figure 8.1 Signal-to-noise ratio for Raman amplifiers based on forward stimulated scattering.

show that a Raman amplifier based on forward stimulated scattering—a Raman FS amplifier—can be convenient for practical applications.

As for stimulated backward scattering, the number of photons at point $z = 0$ of a given Stokes mode in an optical fiber with length L is given by (6.26), that is,

$$N_{sp}(0) = \int_0^L \alpha S_p(0)\exp\{A'[1 - \exp(-\gamma\xi)] - 2\gamma\xi\}d\xi \qquad (8.29)$$

where A' is now related to the backward Raman net gain coefficient and has the identical mathematical form as the forward Raman amplification coefficient, (8.25), but another numerical value. (The Raman amplification coefficient α' concerning backward scattering is not quite identical to the Raman amplification coefficient α concerning forward Raman scattering.) But the approximations $\alpha \approx \alpha'$ (or $\alpha_0 \approx \alpha_0'$) and $A \approx A'$ are quite justified. Thus, (8.29) can be expressed in the closed form

$$N_{sp}(0) = \left[\frac{1}{A} + \exp(-\gamma L)\right] \exp\{A[1 - \exp(-\gamma L)]\} - \left(\frac{1}{A} + 1\right) \qquad (8.30)$$

The total Stokes power for backward spontaneous Raman scattering can be evaluated under the same assumptions as for forward scattering. Thus, it can be assumed that the loss coefficients for the pump wave and the Stokes wave are identical and that the pump signal attenuation is described by the relation $S_p(z) = S_p(0)\exp(-\gamma z)$. The total power of the backward signal due to spontaneous Raman scattering at the input end of the optical fiber is given by the known equation

$$P_{sp}(0) = \int_{f_p}^{\infty} hfN_{sp}(0)df \tag{8.31}$$

The evaluation of (8.31) can be made assuming that the Raman amplification coefficient, $\alpha' \approx \alpha$ has the constant value $\alpha_0' \approx \alpha_0$ in the relevant wavelength region $\Delta\lambda$ defined by the amplifier filter. After these assumptions, (8.31) takes the following closed form:

$$P_{sp}(0) = \Lambda\left\{\left[\frac{1}{A} + \exp(-\gamma L)\right]\exp\left\{A[1 - \exp(-\gamma L)]\right\} - \frac{1}{A} - 1\right\} \tag{8.32}$$

Equation (8.32) is simplified when $\gamma L \gg 1$ and becomes

$$P_{sp}(0) = \Lambda\left[\frac{1}{A}\exp(A) - \frac{1}{A} - 1\right] \tag{8.33}$$

Thus, the amplified spontaneously scattered signal in the backward direction is also independent on the distance L.

In the case of backward Raman scattering, there is a simultaneous propagation of the useful signal injected at the point $z = L$ and traveling in the $-z$ direction, with spontaneously amplified light. The useful signal is amplified by backward stimulated scattering, and its power, P_s, at the point $z = 0$ is given as (see Chapter 6)

$$P_s(0) = P_s(L)\exp(-\gamma L + A) \tag{8.34}$$

where $\gamma L \gg 1$. The SNR at the input point for a Raman backward scattering (BS) amplifier can be found from (8.33) and (8.34) and is given by

$$R_{rB} = 10\ell og(r_{rB}) \tag{8.35}$$

where

$$r_{rB} = \frac{P_s(0)}{P_{sp}(0)} = \frac{P_s(L)}{\Lambda} \frac{\exp(-\gamma L + A)}{\left[\frac{1}{A}\exp(A) - \frac{1}{A} - 1\right]} \tag{8.36a}$$

Equation (8.36a) can be simplified if $A \gg 1$ and rewritten as

$$r_{rB} = \frac{A}{\Lambda} P_s(L) \exp(-\gamma L) \qquad (8.36b)$$

Two important conclusions follow from (8.36). First, the SNR is now dependent on the optical line length, L, and, second, the SNR is dependent on the net Raman gain coefficient, A. The term $P_s(L)\exp(-\gamma L)$ is the power at the output point of the optical fiber with length L in the case without the Raman gain. The dependence $R_{rB}(L)$ is shown in Figure 8.2 for these values:

$$
\begin{aligned}
\gamma &= 4.6 \cdot 10^{-2} \text{ km}^{-1} \text{ (corresponding to 0.2 dB/km)};\\
\lambda_0 &= 1.5 \ \mu\text{m};\\
P_s(L) &= 1 \text{ mW (0 dB)};\\
\Delta\lambda &= [1 \text{ nm and } 0.1 \text{ nm}];\\
\alpha_0 &= 4.6 \cdot 10^{-12} \text{ cm/W};\\
P_p(0) &= 250 \text{ mW}
\end{aligned}
$$

Hence, the level of amplified spontaneous Raman scattered light power (or noise) for backward Raman amplification is higher than the level of amplified spontaneous Raman scattered light power for forward amplification. The difference between these levels is in the range of 5 to 10 dB. The same difference exists between the corresponding SNRs. A Raman FS amplifier is favorable for practical applications in lightwave transmission systems. Besides, the SNR in a Raman FS amplifier does not depend on the Raman amplification coefficient but only on the level of the input signal optical

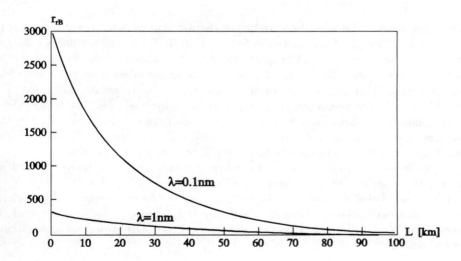

Figure 8.2 Signal-to-noise ratio for Raman BS amplifiers.

power. On the contrary, the SNR in a Raman BS amplifier depends on both the net Raman amplification coefficient and the transmission-line length.

8.3 APPLICATION OF OPTICAL AMPLIFIERS

8.3.1 Application of Laser Amplifiers

8.3.1.1 Optical Preamplifiers

Laser amplifiers can be used as preamplifiers in both lightwave coherent detection systems and lightwave DD systems. There is a difference in these two applications [13, 14], which we will describe in more detail.

The application of a laser amplifier as a preamplifier stage within a DD optical receiver means that an incoming optical signal is first amplified, after that the opto-electronic conversion is made. Thus, the SNR will depend on the same factors as in (8.18). In fact, only the term corresponding to the thermal noise must be included into denominator of (8.18). Thus, the SNR at the output of the photodiode in such a case becomes

$$R_d = \frac{[<N>q]^2}{2q\Sigma^2 B + \dfrac{4k\Theta B}{R}} \tag{8.37}$$

where the mean value and the variance of the incident photons are determined by (8.13) and (8.14). The parameters k, Θ, and R in (8.37) are Boltzmann's constant, the absolute temperature, and the photodiode load resistance, respectively. It is clear that the level of the optical signal power reaching the photodetector is increased by the preamplifier stage, with simultaneous increase of quantum shot noise. The amplification is not accompanied by an increase of thermal noise, so the total SNR at the photodiode output is to be higher than that without optical amplification. But the improvement in optical receiver sensitivity will be rather small, so it can be concluded that this application of laser amplifiers is not very efficient.

If the laser amplifier is used in a coherent optical receiver, the mixing process between the incident optical signal and the local source signal precedes the photodetection process. There are increases in both the mean photon number value and the photon number variance, compared with the DD. Thus, the mean value and the variance of photons in such a case are given as a sum of DD terms and as additional terms, that is,

$$<N>_c = <N>_d + \overline{N}_L + 2\sqrt{G\overline{N}_0\overline{N}_L}\,\cos(2\pi f_c t) \tag{8.38}$$

and

$$\Sigma_c^2 = \Sigma_d^2 + \overline{N}_L + 2\overline{N}_L(G-1)n_{sp} \tag{8.39}$$

where the subscripts c and d mean coherent and direct detection, respectively. Terms $<N>_d$ and Σ_d^2 are determined by (8.13) and (8.14), respectively. In (8.38) and (8.39) n_{sp} means the spontaneous emission factor given by (8.10), $\overline{N}_L$ is the mean number of photons from the local laser source, and f_c is the central frequency of the IF stage. The last term on the right side of (8.39) represents the cross-correlational noise caused by the mixing of spontaneous emitted light in the amplifier and the light signal from the local optical source. Since this term dominates in real situations, the SNR at the output of the IF stage can be expressed as

$$R_c = \frac{2q^2G\overline{N}_0\overline{N}_L}{2q^2[2\overline{N}_L(G-1)n_{sp}]2B} \cong \frac{\overline{N}_0}{4n_{sp}B} \tag{8.40}$$

This ratio is equal to the limit value of the SNR in laser amplifiers. This fact shows that the application of the laser amplifier as a preamplifier in a coherent optical receiver is favorable compared with its application in a DD optical receiver. But we must remember that coherent detection theoretically gives the possibility of reaching the limit value of the SNR caused only by quantum noise of the nonamplified signal. It means that application of the laser amplifier as a preamplifier stage in a coherent optical receiver is not accompanied by an essential improvement of receiver sensitivity. A considerably larger contribution to the optical-receiver sensitivity is observed in the case when laser amplifiers are used as in-line amplifiers.

8.3.1.2 Line Laser Amplifiers

When a laser amplifier is used as an in-line amplifier, the lightwave line is a cascade interconnection of optical-fiber sections and laser amplifiers. Optical signals emitted by the light source in the optical transmitter propagate through this cascade and reach the photodetector in the optical receiver. The main task of one amplifier in such a case is to compensate for optical losses in the optical fiber on the preceding optical section [15].

Some important facts must be mentioned before we consider line optical amplifiers. Due to optical-fiber loss, the total number of photons in the optical signal decreases during the propagation process. Hence, the number of photons reaching the end of a section is lower than the number injected in the optical fiber at the beginning of the section. The decrease in the number of photons is proportional to the factor $\Gamma = \exp(-\gamma L)$, where γ is the loss coefficient, and L is the optical section length. If the mean value and the variance of the photon number at the input end of the optical section are $\overline{N}_0$ and σ_0^2, respectively, the mean value and the variance of the photon number at the output end of the optical section can be expressed, respectively, as

$$<N>_1 = \Gamma\overline{N}_0 \tag{8.41}$$

and

$$\sigma_1^2 = \Gamma\sigma_0^2 \tag{8.42}$$

On the other hand, the mean value and the variance of the photons at the output end of the section with an optical amplifier can be calculated from (8.5) and (8.6) [16–18]. Thus, the following equations are obtained:

$$<N>_2 = \Gamma G\overline{N}_0 + \Gamma(G - 1)n_{sp}\Delta f \tag{8.43}$$

$$\sigma_2^2 = \Gamma G\overline{N}_0 + \Gamma(G - 1)n_{sp}\Delta f + 2\Gamma^2 G(G - 1)n_{sp}\overline{N}_0 + \Gamma^2(G - 1)^2 n_{sp}^2 \Delta f \tag{8.44}$$

The SNR at the output end of the optical section without an optical amplifier is given by

$$R_1 = \frac{<N>_1^2}{\sigma_1^2(2B)} = \Gamma\frac{\overline{N}_0}{2B} \tag{8.45}$$

Hence, the influence of the fiber loss on the optical section is expressed by the factor Γ, where $\Gamma < 1$.

The SNR at the output end of the optical section with an optical amplifier is given as

$$R_2 = \frac{(\Gamma G N_0)^2}{[\Gamma G\overline{N}_0 + \Gamma(G - 1)n_{sp}\Delta f + 2\Gamma^2 G(G - 1)n_{sp}\overline{N}_0 + \Gamma^2(G - 1)^2 n_{sp}^2 \Delta f]2B} \tag{8.46}$$

Because the second and fourth terms in the denominator in (8.46) can be considerably decreased by the proper choice of optical filter following the optical amplifier, these terms can be neglected, and (8.46) takes the form

$$R_2 = \frac{(\Gamma G N_0)^2}{(\Gamma G\overline{N}_0 + 2\Gamma^2 G^2 n_{sp}\overline{N}_0)2B} \tag{8.47}$$

If the considered optical section is the last one on the optical line, the output end of the section is coupled to a photodetector. The SNR at the output of the photodetector is obtained by (8.47), adding the thermal noise term in the denominator. In the case when the thermal noise term dominates, all other terms in the denominator can be neglected, so the SNR is proprotional with Γ^2, which is the worst case that could appear in optical transmission systems with optical amplification by line amplifiers.

When there are a large number of optical amplifiers on the optical line, the total SNR will be different than that given by (8.47). Analysis of such a case would be quite complex without some additional assumptions. It is reasonable to assume that all the optical sections have the same characteristics related to distance length, loss coefficients, and amplification coefficients. Next, it is also reasonable to assume that optical amplifiers only compensate for the optical losses on the line, so it is valid that $G\Gamma = 1$. If the number of optical sections on an optical line is M, the mean value, $<N>$, and the variance, Σ^2, of the photon number at the output end of optical line are, respectively,

$$<N> = \overline{N}_0 + \Gamma(G - 1)n_{sp}M\Delta f \tag{8.48}$$

$$\Sigma^2 = \overline{N}_0 + \Gamma(G - 1)n_{sp}M\Delta f + 2\Gamma(G - 1)n_{sp}M\overline{N}_0 + [\Gamma(G - 1)n_{sp}M]^2\Delta f \tag{8.49}$$

When the thermal noise is very small, the SNR is given by

$$R_{dM} = \frac{q^2N_0^2}{2q^2\sigma^2B} \cong \frac{\overline{N}_0}{4n_{sp}MB} \tag{8.50}$$

The same ratio, but without optical amplifiers, has the form

$$R'_{dM} = \frac{\Gamma^M\overline{N}_0}{2B} = \frac{\overline{N}_0}{G^M(2B)} \tag{8.51}$$

The improvement in the SNR made by optical amplification is measured by the ratio R_{dM}/R'_{dM}, which is equal to the factor $G^M/2Mn_{sp}$.

As for employment of in-line optical amplifiers in a coherent lightwave system, the following equations are valid for the mean value and the variance of the photon number at the output of the transmission line:

$$<N> = \overline{N}_0 + \Gamma(G - 1)n_{sp}\Delta fM + \overline{N}_L + 2\sqrt{\overline{N}_0\overline{N}_L}\ \cos(2\pi f_c t) \tag{8.52}$$

$$\Sigma^2 = \overline{N}_0 + \Gamma(G - 1)n_{sp}\Delta fM + \overline{N}_L + 2\overline{N}_0 n_{sp}M\Gamma(G - 1) \\ + 2\overline{N}_L n_{sp}M\Gamma(G - 1) + \Gamma^2(G - 1)^2 n_{sp}^2 M^2\Delta f \tag{8.53}$$

where $\overline{N}_L$ is the mean number of photons in the local source signal, and f_c is the central frequency of the IF stage. The fifth term on the right side of (8.53) represents the cross-correlational noise caused by the mixing of spontaneously emitted light in the amplifier and the light signal from the local optical source. Since this term dominates in the real situation, the SNR at the output of the IF stage can be expressed as

$$R_{cM} = \frac{2q^2 G \overline{N}_0 \overline{N}_L}{2q^2 [2\overline{N}_L (G-1)n_{sp}]2MB} \cong \frac{\overline{N}_0}{4n_{sp}MB} \qquad (8.54)$$

This is the same value as in the DD case, but it is quite reachable now, because the local oscillator signal suppresses cross-correlational noise due to mutual mixing of the signal and spontaneous emitted light. Thus, the main conclusion is that the application of optical line amplifiers is the most favorable in coherent transmission systems. As for nonlinear systems, it should be pointed out that reshaping rather than amplification is made by laser amplifiers, and some specific phenomena appear. Those phenomena will be analyzed in Chapter 9.

8.3.2 Application of Raman Amplifiers

The nature of the Raman amplification process favors the application of Raman amplifiers in the role of line amplifiers. Hence, we will discuss such an application. The SNR in Raman amplifiers was evaluated in Section 8.2 for cases of forward and backward scattering. It is important to know what happens in the chain of Raman amplifiers on the optical line.

A chain of Raman FS amplifiers means periodical injection of a pump in the direction of information carrying signal propagation. A signal with power $P_s(0)$ is periodically amplified on the optical-fiber line, so the loss influence is periodically compensated. If there are M number of Raman FS amplifiers with distance L between them, the optical signal at the end of the optical line at distance $\ell = ML$, can be expressed as [see (8.26)]

$$P_s(\ell) = P_s(0)A_0 A_0 \ldots A_0 = P_s(0)A_0^M \qquad (8.55)$$

where

$$A_0 = \exp\left(\frac{\alpha_0 S_p}{\gamma} - \gamma L\right) \qquad (8.56)$$

is the normalized net-amplification coefficient. At the same time, it is easy to conclude that the cascade interconnection of several Raman amplifiers will lead to an accumulation of spontaneously scattered light, which presents the noise. According to (8.27), the total noise at the output end of the optical-fiber line can be expressed as the sum

$$P_{sp}(\ell) = \Lambda A_0^M + \Lambda A_0^{M-1} + \ldots + \Lambda A_0^2 + \Lambda A_0 = \Lambda \frac{A_0(1 - A_0^M)}{1 - A_0}, \ (A_0 \neq 1) \quad (8.57)$$

The SNR at the end of the optical line containing M Raman FS amplifiers can be found from (8.55) and (8.57), that is,

$$r_{rF}(\ell) = \frac{P_s(\ell)}{P_{sp}(\ell)} = \frac{p_s(0)}{\Lambda} \frac{A_0^{M-1}(1 - A_0)}{1 - A_0^M} = r_{rF}(L)f_F(A_0, M) \quad (8.58)$$

Therefore, the SNR at the end of the transmission line is expressed by the product of the same ratio that refers to the single repeater section L and a correction term, depending on the amplification coefficient, A_0, and the number of repeater sections. The normalized net-amplification coefficient, A_0, should be chosen to be equal to unity or, more exactly, to be slightly above unity, because of the condition in (8.57).

The SNR for the chain of Raman BS amplifiers can be found in the same way. The signal power at the end of the transmission line consisting of M repeater sections can be expressed as

$$P_s(\ell) = P_s(0)A_0^M \quad (8.59)$$

We must remember that the pump for Raman BS amplifiers must be injected in the $-z$ direction to cause the signal amplification propagating in the z+ direction. Hence, (8.59) is only an adapted one, following from (8.34). Since the scattered amplified light is accumulated over the transmission line with total length ℓ, the total power of this light, meaning the noise at the output end, can be found as

$$P_{sp}(\ell) = P_{sp}(L) + P_{sp}(L)A_0 + \ldots + P_{sp}(L)A_0^{M-2} + P_{sp}(L)A_0^{M-1}$$
$$= P_{sp}(L)\frac{1 - A_0^{M-1}}{1 - A_0}, \ (A_0 \neq 1) \quad (8.60)$$

where $P_{sp}(L)$ has the meaning of $P_{sp}(0)$ in (8.33), or

$$P_{sp}(L) = \Lambda \left[\frac{1}{A} \exp(A) - \frac{1}{A} - 1\right] \quad (8.61)$$

The SNR at the output end of the transmission line consisting of M repeater sections with Raman BS amplifiers is obtained from (8.59) and (8.60), so it becomes

$$r_{rB}(\ell) = \frac{P_s(\ell)}{P_{sp}(\ell)} = \frac{P_s(0)}{\Lambda} \left[\frac{1}{A} \exp(A) - \frac{1}{A} - 1\right]^{-1} \frac{A_0^{M-2}(1 - A_0)}{1 - A_0^{M-1}}$$
$$= r_{rB}(L)f_B(A_0, M) \quad (8.62)$$

Thus, the SNR at the end of the optical-fiber transmission line with Raman BS amplifiers is expressed by the product of the single-section SNR and a correcting term depending on the normalized amplification coefficient and the number of sections. Figure 8.3 shows the correction terms $f_F(A_0, M)$ and $f_B(A_0, M)$.

It is clear that the accumulation of amplified scattered light on a transmission line with a lot of Raman amplifiers will limit the total transmission line length. The total transmission line length, ℓ, is determined by the number of sections and the repeater distance length. It is most suitable to choose that $A_0 = 1 + \epsilon$, where ϵ is an arbitrary small number, so the repeater distance length is given as

$$L = \frac{\alpha_0 S_p(0)}{\gamma^2} \tag{8.63}$$

where $S_p(0)$ now refers to the injected pump power for every Raman amplifier in the chain of amplifiers. Thus, for the values $\alpha_0 = 4.6 \cdot 10^{-12}$ cm/W, $\gamma = 4.6 \cdot 10^{-7}$ cm^{-1} (corresponding to 0.2 dB/km loss coefficient), and $S_p(0) = 250$ mW/(50 μm^2), it is obtained that $L \approx 100$ km.

To evaluate the number of amplifiers in the transmission line we must bear in mind the allowed SNR degradation. The SNR should not be decreased below some critical value. Since the SNR degradation is caused by terms $f_F(A_0, M)$ and $f_B(A_0, M)$, the allowed number of Raman amplifiers can be estimated. In accordance with the

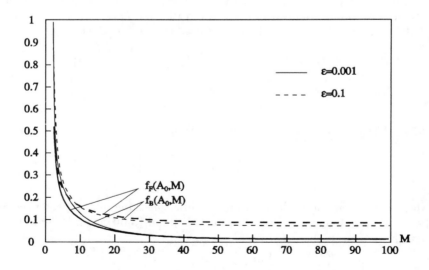

Figure 8.3 The correction terms for signal-to-noise ratio in a Raman optical amplifiers chain.

data from Figures 8.1, 8.2, and 8.3, we can estimate that the allowed number of Raman FS amplifiers on the transmission line is about 50, while the allowed number of Raman BS amplifiers is about 40. Accordingly, we can assume that transmission over a 5000-km optical-fiber line is possible with Raman FS amplifiers, and transmission over 4000 km is possible with Raman BS amplifiers. These lengths are valid for a pump power of about 250 mW and standard single-mode optical fiber with a 50-μm^2 cross-sectional area. The total length of transmission line is directly proportional to the value of pump power. It is reasonable to suppose that it will be possible to achieve 0.5-W pump power, so doubled total transmission line lengths, rather than calculated ones, could be expected. Although this analysis of Raman amplifiers' chain behavior is related to DD lightwave systems, it can be easily concluded that their application in coherent detection lightwave systems is even more efficient due to suppression of the spontaneous scattered amplified light.

8.3.3 Design Characteristics of Optical Amplifiers

There are several important questions concerning the design of optical amplifiers: How to interconnect the amplifiers and optical fiber sections? How to inject the pump signal? How to stabilize pump source characteristics? We will consider some of these effects in the next part for both laser and Raman amplifiers, by using the general scheme of amplifier–optical line interconnection shown in Figure 8.4.

One of the basic problems that appears due to the interconnection of laser amplifiers with optical fiber sections is related to the coupling of the active region (or resonant cavity) with the incoming and outgoing optical fiber. There are some additional losses at the fiber-laser connection area. These losses "neutralize" a part of the gain achieved during signal propagation through the active region and are similar in character to the losses at the connection of the light source with the optical fiber. The most efficient operation of the laser amplifiers occurs in the combination with single-mode optical fiber, which is a favorable fact.

The operation principle of the laser amplifier is basically the same as that of a semiconductor laser. There should be pumping by some bias current, I, which determines the operating point of the laser. The pumping is accompanied by information-signal injection. Thus, in effect the bias current is a basic pump, and a weak optical signal is a modulation pump. The total effect of such a scheme is operation above the lasing threshold and stimulated emission of the light. It is important to stabilize the laser operating regime by the same methods as stabilization of semiconductor laser sources. Operation of the laser amplifier is equivalent to operation of the laser source, because there are bias currents in both operating schemes and modulation signals as well. The main difference is that the modulation signal in the laser source has an electrical nature, while the modulation signal in the laser amplifier has an optical nature.

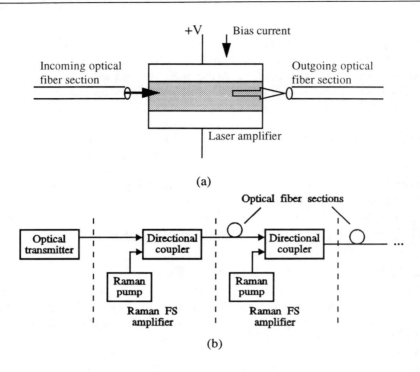

Figure 8.4 The amplifier-optical line interconnection: (a) laser amplifier, (b) Raman amplifier.

The maximal effects of amplification are observed when an incoming optical signal to the laser amplifier possesses only one polarization state. In addition, the total amplification effects will depend on the quality of facet polishing and of the antireflection coating. These two practical aspects determine both the reflection of the incoming optical signal from the face facet and the resonant quality of the active region.

The effective gain achieved in a laser amplifier is, in general, about 30 dB, when the bias current is very close to the threshold current. Since the power-current characteristics of the laser structure do not have a linear character, the amplification coefficient will depend on the incoming optical signal level as well. If the incoming signal power increases, the amplifier can reach the saturation regime, where the amplification effect decreases [17]. It may play an important role in certain circumstances, such as application in multiterminal systems, where the distances between terminals are not equivalent, and incoming signals from the nearest terminal may have a relatively high level at the amplifier input. Such situations require an additional optimization procedure concerning the pump amplifiers and laser resonator design.

The described cascade interconnections of laser amplifiers and optical fiber sections is the common manner for practical amplification of the optical signal or the transmission line [2, 4, 18]. But other solutions, such as that described in [19], also have been proposed. The basic idea of the laser-amplifier realization from [19] is a "parallel" interconnection between laser and optical fiber instead of their serial connection. That interconnection is made by removing the cladding of the optical fiber over the defined length and "gluing" the dye laser instead of the cladding. The pumping of the dye laser is made by a Nd:YaG laser through an optical coupler placed before the dye laser. This operating principle eliminates the current pump and introduces so-called *active optical fiber media*. The concept of active optical fiber media is very attractive, and we will analyze such a concept in the example of erbium-doped optical-fiber amplifiers.

Experimental realizations of Raman amplifiers have showed that it is possible to obtain a net gain above 15 dB, [12, 20]. The noise level in the experimental realization is in the theoretically predicted region [20].

Although we discuss only Raman amplifiers, it should be noted that Brillouin amplifiers also can be efficiently applied in certain circumstances for amplification of optical signals. The efficiency of such amplifiers is increased by the spreading of their amplification bandwidth. The amplification bandwidth can be spread electronically, as in [21]. The basic idea for bandwidth spreading was in frequency modulation of the pump signal, and bandwidth spread from 15 MHz to 150 MHz was observed. By frequency modulation an increase of the total gain was also achieved. The basic advantage of the Brillouin amplifiers compared with Raman amplifiers is lower pump power for the stimulated emission. That pump power for wavelength 1.3 μm is only 5 mW. Hence, Brillouin amplifiers can be used for amplification of optical signals with relatively small bit rates. The limit value of the bit rate, when the application of Brillouin amplifier is possible, is about 20 Mb/s [21].

8.4 ERBIUM-DOPED OPTICAL-FIBER AMPLIFIERS

Optical amplifiers, in the form of optical fibers doped with suitable ions, have been proposed for many years for signal amplification. It seems that rare-earth ions are the most suitable dopants for that purpose [22] and that erbium (Er^{3+}) ions are the most suitable of those. Thus, erbium probably will be the most often used dopant. One reason for this suitability is found in the erbium ions' luminescence spectrum, which is shown in Figure 8.5 for a silica-based optical fiber. The gain has been observed over almost the entire emission spectrum, but it is over 20 dB in the range approximately between wavelengths 1530 and 1560 nm [23]. Hence, such an active fiber device would operate in the 1550-nm wavelength region, the most suitable region for optical-signal transmission in silica-based optical fibers.

The active fiber device, obtained by incorporation of erbium ions into the glass matrix of the core of a silica optical fiber, is the classical three-level lasing system illustrated in Figure 8.6. Sufficient population inversion is created by transition of

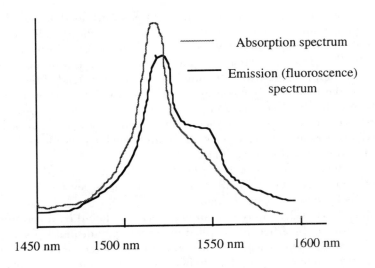

Figure 8.5 Emission and absorption spectra of erbium-doped silica fiber.

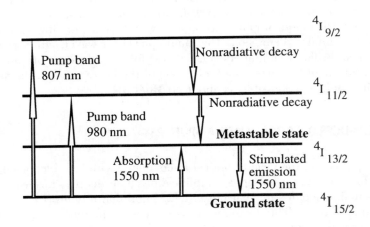

Figure 8.6 Energy levels in erbium-doped fiber amplifier.

electrons from the ground ($^4\text{I}_{15/2}$) level to the top by an optical pumping. The top level can be either $^4\text{I}_{9/2}$ level (created by an 800-nm pump) or $^4\text{I}_{11/2}$ level (created by a 980-nm pump). The fast, nonradiative transitions happen between these two top levels and the metastable medium-energy state $^4\text{I}_{13/2}$, which is the initial level for the radiative transitions back to ground level. The radiative transitions produce gain in the spectrum, which is illustrated in Figure 8.5. Hence, the stimulated emission is observed between the medium and the ground-state energy levels, giving the wavelengths around 1530 nm. If some weak optical signal with the wavelength from this region is present in such a laser structure, the stimulated emission will occur, or amplification of the incident light signal will be made.

The pump wavelength in erbium-doped amplifiers can be injected by an optical coupler. Since the pump wavelength corresponds to the emission of high-power GaAlAs lasers, it is possible to realize pump devices that are essentially compatible with the other elements in a lightwave transmission system. It has been observed that an 800-nm pump is more efficient than a 980-nm pump [22, 23]. It also has been shown [22] that an active erbium-doped optical-fiber length should be only several meters. In such a structure, an amplification coefficient up to 20 dB can be obtained, for pump powers of about 40 mW. There is, however, a certain temperature instability of the amplification coefficient, and that inconvenience should be removed or minimized.

To find the maximum gain and the noise characteristics of erbium-doped optical amplifiers, the rate equation describing the effects of the pump, P_p, the radiated signal, P_s, and the amplified spontaneous emission, P_a, powers should be established. Sometimes the analysis of noise characteristics of erbium-doped optical-fiber amplifiers can be simplified if we keep in mind that short-length optical-fiber amplifiers (or lumped amplifiers) are, in fact, a kind of traveling-wave amplifier, while long-length optical-fiber amplifiers (or distributed amplifiers) can be treated as a special kind of stimulated scattering-based amplifier. Hence, the obtained results for laser amplifiers and Raman amplifiers can be applied for optical-fiber amplifiers as the first approximation, depending on the type of optical fiber amplifier (lumped or distributed).

There is no doubt that erbium-doped optical-fiber amplifiers will find future application in both coherent and nonlinear lightwave communications systems [23–28] as optical in-line amplifiers rather than as preamplifiers. Such an amplifier has several important advantages over other types of optical amplifiers, such as the most suitable wavelength and relatively low pump power and geometrical compatibility, so it can be a form-scheme for the future design of optical amplifiers. Next, we will analyze the characteristics of erbium-doped fiber amplifiers in more detail.

8.4.1 Characteristics of Erbium-Doped Fiber Amplifiers

Research on rare earth–doped optical fibers leapt to prominence with the development of high-gain erbium-doped fiber amplifiers for the important wavelength region around 1550 nm. Some of the notable results achieved with erbium-doped fiber amplifiers are over 50-dB optical gain, 20-dBm saturated optical power, and 3-dB noise figure [29]. Because such high-performance optical amplifiers may enhance both coherent and IM/DD lightwave systems, they have been intensively used in various experiments [30, 31].

To analyze the main characteristics of erbium-doped fiber amplifiers, we will use the three-level system presented in Figure 8.6. Levels 1 and 2 correspond, respectively, to the $^4I_{13/2}$ and $^4I_{15/2}$ levels of erbium ions, while level 3 corresponds to the pump absorption band. The rate equations for the two first-level populations, N_1 and N_2, can be written in the form [32]

$$\frac{dN_1}{dt} = -(W_{12} + R)N_1 + (W_{21} + A_{21})N_2 + RN_3 \tag{8.64}$$

$$\frac{dN_2}{dt} = W_{12}N_1 - (W_{21} + A_{21})N_2 + A_{32}N_3 \tag{8.65}$$

The difference $N_2 - N_1$ corresponds to the population-inversion-generating peak wavelength in the emission spectrum, while A_{21} and A_{32} are the spontaneous decay rates from level 2 to level 1 and from level 3 to level 2, respectively. The inversion values of these rates present characteristic lifetimes ($\tau_{21} = 1/A_{21}$ is the so-called fluorescence lifetime). Parameters W_{12} and W_{21} are the stimulated emission rates, while R represents the stimulated pumping rate. Stimulated emission rates and the stimulated pumping rate depend on the optical pump power, the incoming optical signal power, the spontaneous emission power, and so-called "overlapping parameters" [29]. For simplicity it is sometimes assumed that $W_{12} = W_{21} = W$ and $N_1 + N_2 + N_3 = N_t = 1$.

The emission spectrum, $\alpha(\lambda)$, and the absorption spectrum, $\gamma(\lambda)$, from Figure 8.5 can be presented as [29]

$$\alpha(\lambda) = \sigma_e(\lambda)\Gamma(\lambda)n_{er} \tag{8.66}$$

$$\gamma(\lambda) = \sigma_a(\lambda)\Gamma(\lambda)n_{er} \tag{8.67}$$

where $\sigma_e(\lambda)$ and $\sigma_a(\lambda)$ are the emission and absorption cross-sections, respectively, $\Gamma(\lambda)$ is the overlap integral between the optical mode and the erbium ions, and n_{er} is the density of the erbium ions. It is rather difficult to accurately measure the functions $\sigma_e(\lambda)$ and $\sigma_a(\lambda)$ in characterizing erbium-doped optical fibers. Because of that, the values of $\sigma_e(\lambda)$ and $\sigma_a(\lambda)$ are often calculated indirectly with the Fuchtbauer-Landenberg equation [33].

The convective equation describing the spatial development of the pumping signal, the radiated signal, and amplified spontaneous emission in erbium-doped fiber amplifiers are given as [34]

$$\frac{dP_p^{\pm}(z, t)}{dz} = \pm P_p^{\pm}\Gamma_p(\sigma_{pa}N_1 - \sigma_{pe2}N_2 - \sigma_{pe}N_3) \pm \gamma_p P_p^{\pm} \tag{8.68}$$

$$\frac{dP_s(z, t)}{dz} = P_s\Gamma_s(\sigma_{se}N_2 - \sigma_{sa}N_1) - \gamma_sP_s \tag{8.69}$$

$$\frac{dP_a^\pm(z, t)}{dz} = \pm P_a^\pm\Gamma_s(\sigma_{se}N_2 - \sigma_{sa}N_1) \pm 2\sigma_{se}N_2\Gamma_s h\nu_s\Delta\nu \pm \gamma_sP_a^\pm \tag{8.70}$$

A superscript plus sign (+) in relations (8.68) to (8.70) denotes pump and amplified spontaneous emission copropagating with the signal, while a superscript minus sign (−) denotes the case when amplified spontaneous emission counterpropagates to the signal. Subscripts a, e, p, and s refer to the absorption, emission, pump, and signal, respectively. The signal to the core overlap is designed by Γ_s, while the pump to the core overlap is Γ_p.

The second term in (8.70) is amplified spontaneous emission power produced in amplifier per unit length in the optical frequency bandwidth, $\Delta\nu$. The loss coefficients γ_s and γ_p represent internal loss of the amplifier, which is particularly important for the case of a distributed optical amplifier where these coefficients are the signal and pump attenuation in the optical fiber.

Equations (8.64), (8.65), and (8.68) to (8.70) should be solved for steady-state conditions (for $dN_1/dt = 0$ and $dN_2/dt = 0$) in order to evaluate the amplified signal power and the output amplified spontaneous emission. The amplifier gain is given as

$$G = \frac{P_s(L)}{P_s(0)} \tag{8.71}$$

while spontaneous emission factors in the forward and backward directions are

$$n_{sp}^+ = \frac{P_a^+(L)}{2h\nu_s\Delta\nu(G - 1)} \tag{8.72}$$

$$n_{sp}^- = \frac{P_a^-(L)}{2h\nu_s\Delta\nu(G - 1)} \tag{8.73}$$

Parameter L represents a normalized length of the amplifier and is given as

$$L(\lambda_s) = L_A\Gamma_s\sigma_{se}(\lambda_s)(N_1 + N_2 + N_3) \tag{8.74}$$

where L_A is the physical length of the erbium-doped fiber amplifier.

Simplification of equations (8.71) to (8.73) leads to the convenient expressions for the amplifier gain, G, and noise figure $N_f = 2n_{sp}$. If $P_s = 0$ and the pump power level is high enough, the following expression for the achievable gain and the achievable noise figure can be obtained [34]:

$$G_{max}(L_{ref}, \lambda_p, \lambda_s) = \exp\left(\frac{\sigma_{se}(\lambda_s)}{\sigma_{se}(\lambda_{ref})} \frac{r_p(\lambda_p) - r(\lambda_s)}{1 + r_p(\lambda_p)} L_{ref} \right) \qquad (8.75)$$

$$N_{fmin}(\lambda_p, \lambda_s) = \frac{2r_p(\lambda_p)}{r_p(\lambda_p) - r(\lambda_s)} \text{ for } L \gg 1 \qquad (8.76)$$

where L_{ref} represents normalized amplifier length at some reference signal wavelength, λ_{ref}. Parameters $r_p = \sigma_{pa}/\sigma_{pe}$ and $r = \sigma_{sa}/\sigma_{se}$ can be easily determined if we know the values of the emission and absorption cross-sections. Emission and absorption cross-sections are determined in [29] and [33], and for wavelengths $\lambda_p = 1480$ nm and $\lambda_s = 1545$ nm, they are

$$\sigma_{pa} = 1.86 \cdot 10^{-21} \text{ cm}^2;$$
$$\sigma_{pe} = 0.42 \cdot 10^{-21} \text{ cm}^2;$$
$$\sigma_{sa} = 2.85 \cdot 10^{-21} \text{ cm}^2;$$
$$\sigma_{se} = 5.03 \cdot 10^{-21} \text{ cm}^2.$$

It leads to the values $r_p = 4.428$ and $r = 0.567$.

According to (8.75) and (8.76), the upper bound of the amplifier gain and the lower bound of the spontaneous emission noise figure can be calculated for various optical wavelengths and amplifier lengths. It was calculated in [34] that, for $L_{ref} = 9$ and $\lambda_{ref} = 1531$ nm, maximum gain, G_{max}, calculated for peak signal wavelength, varies from $G_{max} = 10$ dB for pump wavelength $\lambda_p = 1510$ nm to $G_{max} = 40$ dB for $\lambda_p = 1450$ nm. At the same time, the noise figure varies from about 6 dB for pump wavelength $\lambda_p = 1510$ nm, down slightly above 3 dB for $\lambda_p = 1480$ nm. It means that quantum-limited signal amplification is practically achieved for the pump wavelength $\lambda_p = 1450$ nm. The results apply to alumino-silicate host optical fiber.

8.4.2 Concatenated In-Line Erbium-Doped Fiber Amplifiers

In a lightwave transmission system (coherent or IM/DD), the optical amplifiers are concatenated with the optical-fiber sections, and optical filters are usually inserted to suppress amplified spontaneous emission noise outside the optical signal band, as illustrated in Figure 8.7. The gain saturation of the optical amplifiers forming a chain is caused by the optical power leaking to amplified spontaneous emission and by signal and noise amplification originating from previous stages.

The saturated gain of the erbium-doped fiber amplifier can be expressed in an implicit form as [34–35]

$$G^{sat} = G \exp\left[(1 - G^{sat}) \frac{P_{in}}{P_{sat}} \right] \qquad (8.77)$$

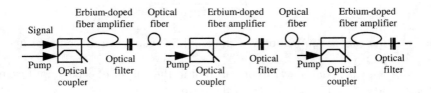

Figure 8.7 Concatenated erbium-doped fiber amplifiers.

where G denotes the amplifier gain in the unsaturated regime, P_{in} is the total optical power at the input of the erbium-doped fiber amplifier, and P_{sat} denotes the saturation power level. The total output power from the amplifier will be about $0.7\,P_{sat}$ if the saturated gain is half the unsaturated gain. The total optical power (signal plus amplified spontaneous emission noise) at the output of the ith amplifier in the chain is

$$P_i^\Sigma = \Gamma\, G_i^{sat} P_{i-1}^\Sigma + 2n_{sp}(G_i^{sat} - 1)h\nu_s B_s, \quad i = 1, 2, \ldots, M \qquad (8.78)$$

where Γ is the optical loss at the amplifier section, G_i^{sat} is the gain of ith amplifier, and B_s is the optical-filter bandwidth. Optical-signal power at the output of the ith amplifier is

$$P_i^s = \Gamma G_i^{sat} P_{i-1}^s \qquad (8.79)$$

The unsaturated gain and saturation power depend on the amplifier design. The saturation power can be regulated by changing the pump power. Unsaturated gain can be kept constant and the saturation power adjusted to reach the most favorable system characteristics. Or, with the same aim, the saturation power can be constant with the variable unsaturated gain. In a chain of in-line erbium-doped optical amplifiers, these variants can be combined to improve the system characteristics [35].

If the total optical power is kept constant during the transmission, or if $P_i^\Sigma = P_o^s$ (where P_o^s corrsponds to the signal level at the output of the optical transmitter), the saturated gain obtained from (8.78) and (8.79) is

$$G^{sat} = \frac{P_o^s + 2n_{sp}h\nu_s B_s}{2n_{sp}h\nu_s B_s + \Gamma P_o^s} \qquad (8.80)$$

Since the output optical power, P_o^s, is much higher than the spontaneous emission noise power represented by $2n_{sp}h\nu_s B_s$, the saturated optical power can be expressed as

$$P_{sat} = \frac{(1 - \Gamma)P_o^s}{\ell n(\Gamma G)} \quad (8.81)$$

This relation can be applied only if the amplifier gain entirely compensates optical loss, that is, if $G^{sat} = 1/\Gamma$.

For optical signal power to be constant at the output of each amplifier, or with $P_i^s = P_o^s$ and $G_i^{sat} = 1/\Gamma$, (8.78) takes the form

$$P_i^{\Sigma} = P_o^s + \quad i \quad 2n_{sp}(G_i^{sat} - 1)h\nu_s B_s \quad (8.82)$$

Now the following expression for P_i^{sat} can be obtained from (8.77) and (8.82):

$$P_i^{sat} = \frac{(1 - \Gamma)}{\ell n(\Gamma G)}[P_o^s + \quad i \quad 2n_{sp}(1/\Gamma - 1)h\nu_s B_s] \quad (8.83)$$

This method can be the preferred method of practical operation, although it is not the most efficient one. It was concluded in [34] that system operation, supervision, and maintenance, rather than noise, will dictate what is the most appropriate method for the design of lightwave transmissions system with a chain of erbium-doped fiber amplifiers.

8.4.3 Some Design Aspects of Erbium-Doped Fiber Amplifiers

As already mentioned in this section, two types of long-distance transmission methods employ erbium-doped fiber amplifiers. One is the lumped-amplifier technique, the other one the distributed-amplifier technique. In the latter method, the whole transmission optical fiber is erbium doped.

The major difference between the lumped and the distributed method is the energy excursions. In a lightwave transmission system with a lumped gain, the path average signal power is different from the power at the output of each optical amplifier and from the optical power of the launched signal. Because of that, in a nonlinear lightwave system, the power of the launched signal should be adjusted to give a path power equal to the soliton power [36] (see Section 9.2.2.5).

In a lightwave transmission system based on distributed amplification, the launched optical power is identical to the path-averaged power if erbium-doped fiber is pumped bidirectionally and there is an optimization in the concentration of the erbium ions in a refractive-index optimized optical fiber [37]. The energy excursions can be kept within ±2 dB range in a 100-km-long erbium-doped optical fiber. Distributed amplification is preferable as long as the change in signal energy over an infinitesimal optical fiber length is smaller in distributed erbium-doped fiber compared to the exponential decay in a lumped-gain transmission link [38].

REFERENCES

[1] Mukai, T., et al., "S/N and Error Rate Performance in Semiconductor Laser Preamplifier and Linear Repeater Systems," *IEEE J. Quantum Electron.*, QE-18(1982), pp. 1560–1568.

[2] Brosson, P., "Analytical Model of Semiconductor Optical Amplifier," *IEEE/OSA J. Lightwave Techn.*, LT-12(1994), pp. 49–54.

[3] Simon, J. C., "Polarization Characteristics of a Traveling Wave Type Semiconductor Laser Amplifier," *Electron. Lett.*, 18(1982), pp. 438–439.

[4] Eisenstein, G., et al., "Gain Measurements of InGaAsP 1.5 μm Optical Amplifiers," *Electron. Lett.*, 21(1985), pp. 1076–1077.

[5] Nakazave, M., et al., "Raman Amplification in 1.4–1.5 μm Spectral Region in Polarization Preserving Optical Fibers," *J. Opt. Soc. America B*, 2(1985), pp. 515–521.

[6] Hegarty, J., et al., "CW Pumped Raman Preamplifier in a 45 km Long Fiber Transmission System Operating at 1.5 μm and 1 Gb/s," *Electron. Lett.*, 21(1985), pp. 290–292.

[7] Olsson, N. A., and J. P. van der Ziel, "Fiber Brillouin Amplifier with Electronically Controlled Bandwidth," *Proc., 9th Conf. Optical Fiber Commun.*, Atlanta, 1986, paper P06.

[8] Anslie, B. J., "A Review of the Fabrication and Properties of Erbium-Doped Fibers for Optical Amplifiers," *IEEE/OSA J. Lightwave Technol.*, LT-9(1991), pp. 220–227.

[9] Shimada, K., et al., "Fluctuation in Amplification of Quanta Width Application of Maser Amplifiers," *J. Phys. Soc. Japan*, 12(1957), pp. 686–700.

[10] Yamamoto, Y., "AM and FM Quantum Noise in Semiconductor Lasers. Part I: Theoretical Analysis," *IEEE J. Quantum Electron.*, QE-19(1983), pp. 34–46.

[11] Mukai, T., and Y. Yamamoto, "Gain, Frequency Bandwidth, and Saturation Output Power of AlGaAs DH Laser Amplifiers," *IEEE J. Quantum Electron.*, QE-17(1981), pp. 1028–1034.

[12] Mochizuki, K., et al., "Amplified Spontaneous Raman Scattering in Fiber Raman Amplifiers," *IEEE/OSA J. Lightwave Techn.*, LT-4(1986), pp. 1328–1333.

[13] Yamamoto, Y., and H. Tsuchiya, "Optical Receiver Sensitivity Improvement by a Semiconductor Laser Preamplifier," *Electron. Lett.*, 16(1980), pp. 233–235.

[14] Duthuus, T., et al., "Detailed Dynamics Model for Semiconductor Optical Amplifiers and Their Crosstalk and Intemodulation Distortion," *IEEE/OSA J. Lightwave Techn.*, LT-10(1992), pp. 1056–1065.

[15] Simon, J. C., "Semiconductor Laser Amplifier for Single Mode Optical Fiber Communication," *J. Optical Commun.*, 4(1983), pp. 51–62.

[16] Okoshi, T., and K. Kikuchi, *Coherent Optical Communications*, Tokyo: Kluwer Academic Publishers, 1988.

[17] O'Mahoni, M. J., "Semiconductor Laser Amplifiers as Repeaters," *Proc., 11th European Conference on Optical Commun.*, Venice, 1985.

[18] Yamamoto, Y., "Noise and Error Rate Performance of Semiconductor Laser Amplifiers in PCM-IM Optical Transmission," *IEEE J. Quantum Electron.*, QE-16(1980), pp. 1073–1081.

[19] Sorin, W. V., et al., "Evanescent Amplification in Single Mode Optical Fibre," *Electron. Lett.*, 19(1983), pp. 820–821.

[20] Olsson, N. A., and J. Hegarty, "Noise Properties of a Raman Amplifier," *IEEE/OSA J. Lightwave Tech.*, LT-4(1986), pp. 396–399.

[21] Olsson, N. A., and J. P. Van der Ziel, "Fibre Brillouin Amplifier With Electronically Controlled Bandwidth," *Electron. Lett.*, 22(1986), pp. 488–490.

[22] Miller, C. A., et al., "Thermal Properties of an Erbium-Doped Fibre Amplifier," *IEE Proc. Pt. J.*, 137(1990), pp. 155–162.

[23] Saito, S., et al., "An Over 2200 km Coherent Transmission Experiment at 2.5 Gb/s Using Erbium-Doped Fiber In-Line Amplifiers," *IEEE/OSA J. Lightwave Techn.*, LT-9(1991), pp. 161–169.

[24] Walker, G. R., et al., "Erbium-Doped Fiber Amplifier Cascade for Multichannel Coherent Optical Transmission," *IEEE/OSA J. Lightwave Techn.*, LT-9(1991), pp. 182–193.

[25] Nakagawa, K., et al., "Trunk and Distribution Network Application of Erbium-Doped Fiber Amplifier," *IEEE/OSA J. Lightwave Techn.*, LT-9(1991), pp. 198–208.

[26] Mollenauer, L. F., et al., "Long-Distance Soliton Propagation Using Lumped Amplifiers and Dispersion-Shifted Fibers," *IEEE/OSA J. Lightwave Techn.*, LT-9(1991), pp. 194–197.

[27] Taga, H., et al., "Multi-Thousand Kilometer Optical Soliton Data Transmission Experiments at 5 Gb/s Using an Electroabsorption Modulator Pulse Generator," *IEEE/OSA J. Lightwave Techn.*, LT-12(1994), pp. 231–235.

[28] Giles, C. R., et al., "2-Gbit/s Signal Amplification at $\lambda = 1.53$ μm in an Erbium Doped Single Mode Fiber Amplifier," *IEEE/OSA J. Lightwave Tech.*, LT-7(1989), pp. 651–656.

[29] Giles, C. R., and E. Desurvire, "Modeling Erbium Doped Fiber Amplifiers," *IEEE/OSA J. Lightwave Techn.*, LT-9(1991), pp. 271–283.

[30] Taga, H., et al., "Long Distance Multichannel WDM Transmission Experiments Using Er-Doped Fiber Amplifiers," *IEEE/OSA J. Lightwave Techn.*, LT-12(1994), pp. 1448–1453.

[31] Hayashi, Y., et al., "Estimated Performance of 2.488 Gb/s CPFSK Optical Nonrepeated Transmission System Employing High Output Power EDFA Boosters," *IEEE/OSA J. Lightwave Techn.*, LT-11(1993), pp. 1369–1376.

[32] Kikushima, K., "AC and DC Gain Tilt of Erbium Doped Fiber Amplifiers," *IEEE/OSA J. Lightwave Techn.*, LT-12(1994), pp. 463–470.

[33] Barnes, W. L., et al., "Absorption and Emission Cross Section of Er^{3+} Doped Silica Fibers," *IEEE/OSA J. Quantum Electron.*, QE-27(1991), pp. 1004–1009.

[34] Giles, C. R., and E. Desurvire, "Propagation of Signal and Noise in Concatenated Erbium Doped Fiber Optical Amplifiers," *IEEE/OSA J. Lightwave Techn.*, LT-9(1991), pp. 147–154.

[35] Bertilsson, K., and P. A. Andrekson, "Modeling of Noise in Erbium Doped Fiber Amplifiers in Saturated Regime," *IEEE/OSA J. Lightwave Techn.*, LT-12(1994), pp. 1198–1206.

[36] Gordon, J. P., and L. F. Molenauer, "Effects of Nonlinearities and Amplifier Spacing on Ultra Long Distance Transmission," *IEEE/OSA J. Lightwave Techn.*, LT-9(1991), pp. 170–173.

[37] Rottwit, K., et al. "Transmission," *IEEE/OSA J. Lightwave Techn.*, LT-11(1992), pp. 1544–1552.

[38] Rottwit, K., et al., "Long Distance Transmission Through Distributed Erbium Doped Fibers," *IEEE/OSA J. Lightwave Techn.*, LT-11(1994), pp. 2105–2114.

Nonlinear Lightwave Systems

9.1 INTRODUCTION

Nonlinear effects that may occur in optical fibers were considered in Chapter 6 and Chapter 7, where it was pointed out that their influence on lightwave system characteristics may be either favorable (compensation of dispersion effects, signal amplification) or unfavorable (increase of signal loss, cross-talking between channels). The favorable influence is related, in fact, to the possibility of soliton generation and soliton regime preservation in an optical fiber. The primary aim of this chapter is to describe the circumstances that favor soliton generation and the preservation of the soliton regime in an optical fiber. The unfavorable effects will be considered primarily through analysis of the signal cross-talking in the multichannel optical transmission system. The nature of additional losses in a transmission channel due to stimulated Brillouin backscattering already has been analyzed, and there is no need to do it in more detail.

In accordance with these facts, the nonlinear effects concerning the nonlinear lightwave system from Figure 1.2 will be analyzed. First, the conditions for soliton generation will be considered, and then the methods for soliton regime preservation over the optical fiber line will be described. After that, an analysis of signal cross-talking between individual channels in a WDM lightwave system will be made. Next, the BER in a nonlinear lightwave system will be considered, and the methods for its evaluation presented. Finally, some design issues concerning multichannel systems will be described.

9.2 CONDITIONS FOR SOLITON GENERATION AND SOLITON REGIME PRESERVATION

9.2.1 Soliton Generation

The soliton pulses propagating through an optical fiber are generated in a so-called *soliton laser*. A soliton laser generates not only the necessary level of optical power, but the desired shape of soliton envelope, as well. The realization of soliton lasers is still in the experimental phase.

The first experimental results, obtained by Mollenauer and Stolen, were very convincing [1, 2]. The soliton laser was realized on the base of a color-center laser with proper mode locking and frequency stabilization. The principles of a soliton laser are illustrated in Figure 9.1

The laser in Figure 9.1 consists of a mode-locked color-center laser, a pump, and two resonant cavities (a main cavity and a control cavity). The control cavity contains a single-mode polarization-preserving optical fiber with prescribed length L. The resonant cavities are coupled through the mirror M_0 with typical reflectivity of about 70%. Lenses L_1 and L_2 and mirrors M_1, M_2, and M_3 are the auxiliary elements for efficient coupling and reflection of the light ray. The input end of the single-mode fiber and the mirror M_0 are mounted on a translation stage and have movements Δl_1 and Δl_2, respectively. Instead of the movement of the input end (with lens L_1), mirror M_3 can be moved. The translation of all these mentioned elements have the purpose of facilitating final adjustment of the optical path length in the optical fiber arm to be an integral multiple of the main cavity length.

The main cavity belongs to the $Tl^0(1)$-KCl color-center laser, which is continuously tunable over a broadband in the $\lambda = 1.5$ μm wavelength region. The color-center laser is pumped by an actively mode-locked continuous wave (CW) Nd:YaG laser and produces picosecond pulses from the main cavity. The main fraction of pulse energy (around 70% of total energy) travels around the main cavity, but a controlled fraction of pulse energy (around 30% of the total energy) travels around

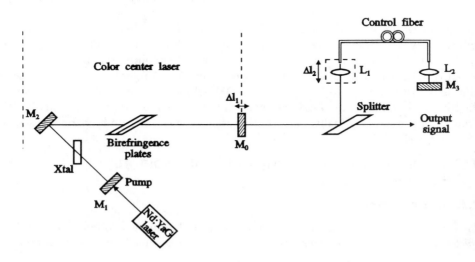

Figure 9.1 Illustration of soliton laser operation. (*After:* [2], reprinted with permission.)

the control cavity. As the laser action builds up from the noise, the relatively broad pulses in a controlled fraction will be considerably compressed in the optical fiber. The compressed pulses are sent back into the main cavity, forcing the color-center laser to produce narrower pulses. This process is successively repeated, and successive narrowing occurs, until the pulses become solitons. At that moment, the stationary state is reached.

Two basic effects influence the stationary state of a soliton laser. First, the time average power inside the control cavity varies with the lens L_1 (or mirror M_2) motion. Since the more efficient work of the soliton laser is with M_2 motion [2], we will follow such a laser scheme. Hence, the power in the control cavity varies with the round-trip optical phase shift φ in the control cavity. Second, the soliton regime of the laser is correlated with some critical optical power, the so-called soliton level power, and that power is within the range of control cavity power.

The dependence of the control cavity power on the optical phase shift φ is the key fact for the analysis of soliton generation and soliton regime stabilization. It has already been mentioned that significant narrowing of the pulse returned from the control fiber is present. At the same time, in a soliton regime, a significant phase retardation is produced by the index nonlinearity in the control optical fiber. To consider both pulse narrowing and phase retardation, the normalized envelope dot product of the signals reaching to the mirror M_0 from the main cavity and from the control cavity should be found. This normalized dot product is [1]

$$I(\xi_L) = \frac{\int_{-\infty}^{\infty} u^*(0, s)u(\xi_L, s)ds}{\int_{-\infty}^{\infty} |u(0, s)|^2 ds} \tag{9.1}$$

where $u(\xi, s)$ and $u(\xi_L, s)$ are the envelope functions of the pulses coming on the M_0 mirror from the main and control cavities, respectively, while s is a dimensionless time measured in pulse-width units [2]. (The envelope functions $u(\xi, s)$ and $u(\xi_L, s)$ are related to the signals with the phases φ_1 and φ_2, respectively). The parameter ξ is a dimensionless length equal to

$$\xi = \frac{\pi z}{2z_0} \tag{9.2}$$

where z is the length coordinate equal to double the length of the optical fiber; that is, $z = 2L$. The value z_0 is the soliton period, while $\xi_L = \pi L/z_0$.

The phase retardation is, in general, nonuniform across the pulse, so an average phase can be defined by

$$\overline{\varphi} = \text{arctg} \frac{\text{Im}[I(\xi_L)]}{\text{Re}[I(\xi_L)]} \tag{9.3}$$

Thus, if φ_2' is the round-trip phase shift in the optical fiber for a broad pulse with low intensity, it is valid that $\varphi_2 = \varphi_2' + \overline{\varphi}$. It is most suitable to compare the value $\overline{\varphi}$ with the phase value φ_m corresponding to the peak of the pulse

$$\varphi_m = \text{arctg} \frac{\text{Im}[u(\xi_L, 0)]}{\text{Re}[u(\xi_L, 0)]} \tag{9.4}$$

It has been shown concerning phase and amplitude relationships that (9.1) to (9.4) are satisfied for the soliton pulses [1, 2]

$$u(\xi, s) = \text{sech}^2(s)\exp(j\xi/2) \tag{9.5}$$

and

$$u(\xi, s) = \frac{4\exp(j\xi/2)[\cosh(3s) + 3\,\exp(4j\xi)\cosh(s)]}{\cosh(4s) + 4\cosh(2s) + 3\cos(4\xi)} \tag{9.6}$$

The color-center-based soliton laser was the first proposed concept providing transform-limited hyperbolic secant–type pulses. After that, several methods have been proposed and used to obtain soliton pulses. Both mode-locking semiconductor lasers with an external cavity [3] and gain-switched distributed feedback (DFB) lasers [4] with ultranarrow band optical filters have been used for soliton generation. The operation of mode-locked semiconductor lasers is extremely sensitive to the ambient perturbation, because of resonator sensitivity to temperature change and to vibration. On the other hand, the pulses produced by gain-switched DFB soliton lasers are of much poorer spectral quality compared to mode-locked lasers because of the large degree of frequency chirping. Therefore, the conventional methods do not seem to be attractive for practical optical soliton pulse generation.

A new method of generating ultrashort optical pulses using an electroabsorption modulator and DFB laser diode has been proposed in [4]. A simple driving scheme that requires no resonators has been proposed and verified experimentally. The proposed method is based on the application of sinusoidal driving voltage on an electroabsorption modulator, which causes the stable generation of ultrashort optical pulses with the almost hyperbolic secant–squared shape given by (9.5).

The second type of soliton lasers proposed so far are so-called *all-fiber lasers* [5]. This type of soliton lasers, operating on the principle of fiber-doped optical amplifiers, offers compatibility with transmission fibers (no need for a pigtail), compact size, and the possibility of passive mode locking. The passive mode locking is based on the reflection characteristics of the nonlinear amplifying loop mirror. In this soliton laser model, the soliton pulses are in fact generated from the amplified spontaneous emission.

9.2.2 Soliton Regeneration on the Optical Fiber Link

Regeneration of soliton pulses on the optical fiber link means their reshaping, or "substitution of a pulsewidth for an intensity." That is to say, during soliton regeneration, the narrowing process occurs, and the loss influence causing pulse spreading is compensated. At the same time, the solitons' amplitude increases, since the area under the soliton envelope must be constant. Several methods for soliton regeneration have been proposed so far [6–12]; we will analyze the most characteristic examples.

9.2.2.1 Regeneration of Solitons by Periodic Injection of Continuous Waves

It was pointed out in Chapter 7 that optical-fiber loss is the only factor that contributes to the spreading of soliton pulses and, consequently, to deterioration of the pulse quality. (The third-order dispersion term contributes to the pulse spreading, but the contribution is very small.) During that process, a soliton does not change its own shape described by the hyperbolic-secant function. Due to soliton spreading, interference of the neighboring pulses must be prevented by periodical regeneration of the solitons.

Since the basic shape of a soliton pulse is being preserved, the main role of the soliton regenerator is to amplify soliton amplitude (amplitude amplification is accompanied by the pulse compression). Amplitude amplification can be accomplished by usage of various types of optical amplifiers, or it can be realized by periodic injection of continuous waves. These waves must have the same frequency and phase as the soliton wave. The employment of optical amplifiers will be described after the analysis of soliton regeneration by continuous-wave injection.

When a continuous wave with the same frequency and same phase as the soliton carrier wave is injected into the soliton pulse train, the soliton can catch the energy of the continuous wave and form a new soliton pulse with a higher energy and narrower pulsewidth. This effect occurs due to the influence of the Kerr effect leading to the contraction of the pulsewidth proportionally to $|E|^2$, where E is the sum of the amplitudes of the soliton and the continuous wave. That process is analyzed by use of the inverse scattering theory [6, 8]. The process of soliton regeneration by continuous-wave injection can be analyzed by use of the normalized Schroedinger equation given by (7.7), or

$$j\frac{\partial q}{\partial z} + \frac{1}{2}\frac{\partial^2 q}{\partial t^2} + |q|^2 q = 0 \qquad (9.7)$$

where $q(t, z)$ is the function of the soliton envelope, t is the time coordinate, and z is the length coordinate. The terms from the right side, concerning the optical-fiber loss and third-order dispersion are neglected in (9.7).

Application of the inverse scattering theory [8, 9] on the Schroedinger equation shows that the amplitude and the phase of a regenerated soliton pulse that arose from the initial soliton shape $q(t, 0)$ can be determined by the complex eigenvalue ζ. The eigenvalue ζ satisfies the equations for the eigenfunctions ψ_1 and ψ_2

$$\frac{\partial \psi_1}{\partial t} + j\zeta \psi_1 = q\psi_2 \tag{9.8a}$$

$$\frac{\partial \psi_2}{\partial t} - j\zeta \psi_2 = -q^*\psi_1 \tag{9.8b}$$

These equations must be solved for eigenvalue ζ to find the structure of the regenerated soliton for a given initial condition $q(t, 0)$. Function $q(t, 0)$ consists of only one soliton with a continuous wave in an infinite time domain, that is,

$$q(t, 0) = u(t, 0) + f(t, 0) \tag{9.9}$$

where $u(t, 0)$ is the soliton given by function

$$u(t, 0) = \eta \, \mathrm{sech}(\eta t)\exp(j\xi t + j\sigma_0) \tag{9.10}$$

while $f(t, 0)$ presents the continuous wave

$$f(t, 0) = E_0 \exp(j\omega t + j\vartheta) \tag{9.11}$$

The eigenvalue $\zeta = \zeta_0$, corresponding to the soliton alone, is

$$\zeta_0 = -\frac{\xi}{2} + j\frac{\eta}{2} \tag{9.12}$$

The variations of the eigenvalue ζ from the value ζ_0, where $\zeta = \zeta_0 + \Delta\zeta$, can be found by a perturbation expansion for $E_0 \ll \eta$. The eigenfunctions ψ_1 and ψ_2 can be expanded as

$$\psi_j = \psi_{j0} + \psi_{j1}, \ (j = 1, 2) \tag{9.13}$$

so from (9.8a) and (9.8b), it follows that

$$\Delta\zeta = \frac{\int_{-\infty}^{\infty} (f\psi_{20}^2 + f^*\psi_{10}^2)dt}{2j\int_{-\infty}^{\infty} \psi_{10}\psi_{20}dt} \tag{9.14}$$

The eigenfunctions ψ_{10} and ψ_{20} are the bound-state solutions to (9.8a) and (9.8b) having the form

$$\psi_{10} = -\frac{1}{2}\operatorname{sech}(\eta t)\exp\left(-\frac{\eta t}{2} + \frac{j}{2}\xi t + \frac{j}{2}\sigma_0\right) \qquad (9.15a)$$

$$\psi_{20} = \frac{1}{2}\operatorname{sech}(\eta t)\exp\left(\frac{\eta t}{2} - \frac{j}{2}\xi t - \frac{j}{2}\sigma_0\right) \qquad (9.15b)$$

At the same time, the real and imaginary parts of the eigenvalue variation

$$\Delta\zeta = -\frac{\Delta\xi}{2} + j\frac{\Delta\eta}{2} \qquad (9.16)$$

take the forms

$$\Delta\xi = E_0\pi\frac{\omega - \xi}{\eta}\cos(\vartheta - \sigma_0)\operatorname{sech}\left(\frac{\omega - \xi}{2\eta}\pi\right) \qquad (9.17)$$

$$\Delta\eta = E_0\pi\cos(\vartheta - \sigma_0)\operatorname{sech}\left(\frac{\omega - \xi}{2\eta}\pi\right) \qquad (9.18)$$

Obtained results for real and imaginary parts show that the soliton amplitude increases proportionally to πE_0 if $\omega = \xi$ and $\vartheta = \sigma_0$. Hence, the amplitude increase is π times as large as that of the linear superposition. At the same time, the soliton pulsewidth is suppressed by the factor πE. The condition $\omega = \xi$ practically means that the continuous wave must have the same frequency as the soliton carrier, while the condition $\vartheta = \sigma_0$ means that the phases of the continuous wave and the soliton carrier must be equal. Hence, if the continuous wave with amplitude E_0 is periodically injected into the optical fiber at distances where the soliton amplitude decreases by the factor $E_0\pi$, the original soliton shape is recovered.

It must be noted, however, that the continuous wave is not totally absorbed by the soliton pulse. The unabsorbed part of the continuous wave will continue propagation as a linear dispersive wave. This linear dispersive wave does not affect the soliton shape since the amplitude of the wave decreases proportionally with $\exp(-\Gamma z)$, while the soliton amplitude decreases as $\exp(-2\Gamma z)^2$ (Γ is the normalized loss coefficient). Meanwhile, the continuous waves will accumulate if the injection is made many times. Because of that, the unused parts of the continuous waves must be eliminated, as proposed in [6].

Since the soliton phase is continuously shifted during the propagation, the phase at distance L will be

$$\sigma(L) = \frac{1}{2}\int_0^L \eta^2 dz + \sigma_0 \qquad (9.19)$$

The soliton amplitude η in (9.19) depends on the distance z due to loss influence, that is,

$$\eta = q_0 \exp(-2\Gamma z)^2 \qquad (9.20)$$

At the same time, the continuous wave is expected to preserve its initial phase, so there is a possibility for mutual elimination of unused parts of continuous waves. If the next continuous wave with frequency $\omega = \xi$ and phase $\vartheta = \pi + \sigma_0$ is injected at the point where the phase of a new continuous wave is equal to the soliton phase, the old continuous wave will be eliminated. Thus, at the same time, both soliton reshaping and elimination of the unused part of the continuous wave are observed.

The realization of this elimination method can be made in two steps. First, the continuous wave (let it be the AA wave) with the amplitude E_0 should be injected at the point when the soliton amplitude is decreased by the factor $E_0\pi$. The continuous-wave amplitude is much lower than the soliton amplitude. Second, another continuous wave (let it be the BB wave) with the phase opposite to the AA wave phase must be injected at the distance when the soliton carrier phase deviates from the phase of the AA continuous wave by factor π. The amplitude E_0 can be chosen such that

$$E_0\pi = q_0[1 - \exp(-2\Gamma z_\pi) \qquad (9.21)$$

where z_π is determined by the relation $\sigma(z_\pi) = \pi + \sigma_0$. In such a way, the distance between two subsequent regenerators that inject the continuous waves is the same and becomes equal to z_π. By this method, complete regeneration of the solitons but without accumulation of continuous waves is achieved. The complete regeneration of the solitons can also be achieved by injection of m continuous waves per regenerative cycle. In such a case, the distance between two regenerators must be such that the phase condition $\sigma(z) = 2\pi/m + \sigma_0$ is satisfied.

A graphical illustration of soliton regeneration by continuous-wave injection is given in Figure 9.2. The initial soliton shape is spread due to optical-fiber loss, and the solitons tend to overlap each other. Amplification must occur before the overlapping to prevent mutual interaction among the solitons. The amplified solitons have a shape almost identical to the initial one.

We should point out that this method requires sophisticated technologies for detection of phase and frequency of the optical wave. That is a big obstacle for practical employment of this method in the foreseeable future. The favorable characteristics of rare earth–doped optical-fiber amplifiers practically overshadow most of the methods proposed so far for soliton regeneration, including this one.

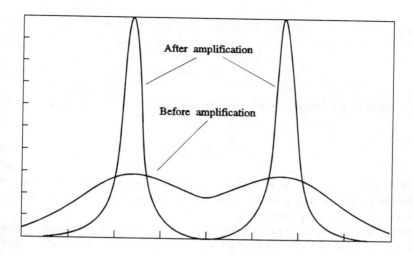

Figure 9.2 Illustration of the soliton amplification in the optical fiber.

9.2.2.2 Soliton Regeneration by Optical Amplifiers

The optical-fiber loss causes a decrease of soliton amplitude and increases the soliton pulsewidth. The amplitude decays as $\exp(-2\Gamma z)$, while the pulsewidth increases proportionally to factor $\exp(2\Gamma z)$. That means that the area under the soliton envelope (approximately equal to the amplitude-pulsewidth product) stays the same even under loss influence. Hence, it can be readily concluded that soliton regeneration can be made by simple amplification without additional reshaping. The amplification can be made by the laser amplifiers described in Chapter 8. If the soliton amplitude is amplified by the factor $\exp(\Gamma z_0)$ at each distance z_0, the effect will be the same as an introduction of the additional term $S(t, z)$ on the right side of (7.30) [6]. The term $S(t, z)$ is given as

$$S(t, z) = j[\exp(\Gamma z_0) - 1]\sum_{n=1}^{N} \delta(z - nz_0)q(t, z) \qquad (9.22)$$

If the condition $\Gamma z_0 \ll 1$ is satisfied, the term $S(t, z)$ becomes

$$S(t, z) = j\Gamma z_0 \sum_{n=1}^{N} \delta(z - nz_0)q(t, z) \simeq j\Gamma \int \delta(x - z)q(t, z)dx \qquad (9.23)$$

So, for $z_0 \to 0$, (9.23) stands for $S(z, t) = j\Gamma q(t, z)$, which leads to the elimination of the loss term in (7.30) and to the exact soliton solution.

During the soliton-amplification process, the amplified pulse has an amplitude that is for a factor b greater than the amplitude of the unamplified soliton, that is,

$$\eta \operatorname{sech}(\eta t) \to b\eta \operatorname{sech}(\eta t), \, b > 1 \qquad (9.24)$$

But the soliton formed from this amplified pulse has the stationary amplitude B given by [6]

$$B = (2b - 1)\eta \qquad (9.25)$$

Then, if $b = \Delta + 1$, it will be $B = (1 + 2\Delta)\eta$; hence, the stationary soliton, created from the amplified soliton pulse, has an amplitude increment twice as large as the increment due to the amplification process.

After the amplification process and formation of the new soliton, a small amplitude dispersive wave is formed as well. The dispersive wave is the rest of the amplified pulse after stationary soliton formation. The energy, ΔE, belonging to the dispersive wave, can be found by subtracting the energy of the stationary soliton from the amplified soliton pulse energy, that is,

$$\Delta E = \int_{-\infty}^{\infty} b^2 \eta^2 \operatorname{sech}^2(\eta t) dt - \int_{-\infty}^{\infty} B^2 \operatorname{sech}^2 (Bt) dt = 2\eta(b - 1)^2 \qquad (9.26)$$

Consequently, in each soliton amplification process, the dispersive wave with the energy given by (9.26) is added.

If the amplification of the solitons is made at shorter distances, the dispersive wave that appears can also be amplified, so it may reach a level sufficient to contribute to form the new soliton. The dispersive wave will be caught and amplified if the distance z_0 between two amplifiers is smaller than some critical length, necessary for the dispersive wave to leave the soliton. It will be discussed in Section 9.2.2.4.

9.2.2.3 Soliton Regeneration by Raman Amplifiers

Soliton amplification by Raman amplifiers seems to be a promising way to realize long-length signal transmission in a nonlinear lightwave system [7, 10]. The basic idea is in periodical pump injection in the optical fiber and induction of the mutual interaction between the pump signal and the soliton's signal. The interaction is achieved by the Raman signal. Since the Raman gain coefficient is relatively small, the peak level of the pump signal should be comparable with the peak power of the soliton pulse. If the pump power is kept constant by adjusting the Raman gain coefficient, so that it compensates the loss coefficient in the optical fiber, the solitons can propagate without distortion at an arbitrary distance along the optical fiber.

However, since the pump power decays with length, the loss influence is compensated only on the average, and either the Raman gain or the loss will dominate at any given position. This effect leads to the periodical soliton perturbations, and the total overlapping of neighboring solitons can occur. Such a state is chaotic and must be prevented. To prevent the mutual interaction between soliton pulses, the ranges of variation of the soliton regime parameters must be established. At the same time, maximal distance between Raman amplifiers must be prescribed for a given bit rate.

To describe the process of soliton amplification by Raman amplifiers, according to the result from [7] and [10], the basic equations from Chapter 6 should be applied to soliton propagation. Thus, the soliton-amplification process can be analyzed by use of the coupled differential equations referring to the solitons (i.e., the scattered signal) and to the pump signal. These differential equations have the form [7]

$$j\left[\frac{\partial E_p}{\partial z} + (\gamma_p + \alpha'|E_s|^2)E_p - k_p'\frac{\partial E_p}{\partial t}\right] - \frac{1}{2}k_p''\frac{\partial^2 E_p}{\partial t^2} + \beta|E_p|^2 E_p = 0 \qquad (9.27a)$$

$$j\left[\frac{\partial E_s}{\partial z} + (\gamma_s - \alpha'|E_p|^2)E_s - k_s'\frac{\partial E_s}{\partial t}\right] - \frac{1}{2}k_s''\frac{\partial^2 E_s}{\partial t^2} + \beta|E_s|^2 E_s = 0 \qquad (9.27b)$$

In (9.27a) and (9.27b), t and z are the time and length coordinates, respectively, while E_p and E_s are the electrical fields of the pump signal and the soliton pulse, respectively. The coefficient α' denotes the coupling coefficient in the Raman process. This coefficient is linked to the self-phase modulation coefficient, χ, ($\chi = \omega n_2/c$; n_2 is Kerr's coefficient, ω is frequency, and c is the light velocity), by the approximate equation [7]

$$\alpha' = 0.2\chi \qquad (9.28)$$

Equations (9.27a) and (9.27b) define the bright solitons if $k'' < 0$, when the energy coupling and fiber loss are absent. It is most suitable to transform (9.27a) in the dimensionless form and then to find its numerical solution. The dimensionless form of (9.27a) is

$$j\frac{\partial q}{\partial \xi} + \frac{1}{2}\frac{\partial^2 q}{\partial \tau^2} + |q|^2 q = -j\Gamma q + j\alpha I q \qquad (9.29)$$

where

$$\xi = 10^{-9}z/\lambda \qquad (9.30)$$

$$\tau = \frac{10^{-4.5}}{(\lambda k'')^{1/2}}\left(t - \frac{z}{v_g}\right) \qquad (9.31)$$

$$q = 10^{4.5}(\pi n_2)^{1/2}E_s \tag{9.32}$$

$$\Gamma = 10^9 \lambda \gamma \tag{9.33}$$

$$\alpha = \alpha'/\chi = 0.2 \tag{9.34}$$

$$I = 10^9(\pi n_2)|E_p|^2 \tag{9.35}$$

(The wavelength is denoted by λ, the group velocity by v_g, while k' and k'' are the first and the second derivative of the wavenumber per frequency ω, respectively.) The normalization of the parameters in (9.29) is made the same way as in (7.7), which concerns the soliton propagation without amplification.

The solution of (9.29) will be bright solitons in the anomalous dispersion region if the Raman amplification coefficient, given by the term αI, is exactly canceled by the loss coefficient Γ. The bright soliton solution has the form

$$q = q_0 \text{sech}(q_0\tau)\exp(jq_0^2\xi/2) \tag{9.36}$$

The nonstationary behavior of the pump intensity, I, during its instantaneous interaction with the soliton can be neglected in the following considerations, and only the stationary space variations can be taken into account. In such a case, the behavior of the pump intensity is described by

$$\frac{dI}{d\xi} = -2\Gamma_p I - 4\alpha I_s I \tag{9.37}$$

where Γ_p is the normalized loss coefficient at the pump signal wavelength, and I_s is the average soliton power given by

$$I_s = \frac{1}{4\Delta t}\int_{-\Delta t}^{\Delta t}|E_s|^2 dt \tag{9.38}$$

where Δt corresponds to the average time period between two solitons.

For small lengths, when $\xi\Gamma \ll 1$, the solution of (9.37) has the form

$$I = I_0 \exp(-G\xi) \tag{9.39}$$

where G is the effective pump loss coefficient, given by

$$G = 2\Gamma_p + 4\alpha I_{s0} \tag{9.40}$$

The values I_0 and I_{s0} in (9.39) and (9.40) are related to the pump intensity and the average soliton power at the point $\xi = 0$, respectively. Now the pump intensity, I_0, which compensates just the losses Γ at the soliton wavelength within the regenerator distance ξ_0, should be determined. It can be assumed that a pump of equal intensity is periodically injected at each distance of ξ_0 in both directions. According to (9.39), the pump intensity will be

$$I = I_0 \exp[-G|\xi - n\xi_0|], \; n = 0, 1, 2, \ldots \qquad (9.41)$$

With such a pump, the integrated gain along the length between the points $\xi = 0$ and $\xi = \xi_0$ stands for

$$g = \alpha \int_0^{\xi_0} \{I_0\exp(-G\xi) + I_0 \exp[G(\xi - \xi_0)]\}d\xi = \frac{2I_0\alpha}{G}[1 - \exp(-G\xi_0)] \quad (9.42)$$

Thus, the value I_0 can be determined so that the integrated gain G just compensates the integrated loss $\Gamma\xi_0$ over the length ξ_0, that is,

$$I_0 = \frac{G\Gamma\xi_0}{2\alpha[1 - \exp(-G\xi_0)]} \qquad (9.43)$$

When the additional loss C_0 (in decibels) is inserted with the pumping, the value of I_0 must be increased to compensate this additional loss. Thus, we have

$$I_0 = \frac{G\Gamma\xi_0 + T}{2\alpha[1 - \exp(-G\xi_0)]} \qquad (9.44)$$

where $T = C_0(\ell n10)/20$.

These results can be used for numerical calculations, such as in [7]. A. Hasegawa has found the numerical solution of (9.29) for typical parameters of single-mode optical fiber and has analyzed some cases that are interesting for practical realization. It has been shown that an initial spacing equal to $\tau_0 = 10\tau$ between neighboring solitons, where τ is initial soliton pulsewidth, is an optimal value for long-length transmission. The distance between two Raman amplifiers has been chosen to be about 20 kilometers. Under these conditions, transmission over 1,000 km is quite possible without significant distortions of the soliton shapes. The proper choices of the initial spacing between neighboring solitons and the distance between two regenerators will prevent mutual interaction among soliton pulses, so the information carried by the soliton pulse train will be preserved.

The influence of two effects must be taken into account in the consideration of Raman amplifier design. The first one is the coupling loss influence, the second is the pump-loss-rate influence. Both of these effects can considerably deteriorate the transmission quality and shorten the transmission line length. The localized coupling loss is injected by a directional optical coupler used for pump power coupling. The coupling losses cause nonadiabatic perturbations in the soliton shape, since the amplitude of the soliton decreases at the coupling point, but this is not accompanied by the soliton pulse spreading. When the amplitude of the soliton is reduced by the factor Δ, the newly formed stationary soliton pulse has the amplitude $(1 - 2\Delta)$ [see (9.25)]. Thus, the $\mathrm{sech}(\tau)$ soliton shape is transformed to a $(1 - 2\Delta)\mathrm{sech}[(1 - 2\Delta)\tau]$ shape. The soliton amplitude is additionally reduced, while the soliton pulsewidth is increased by a factor of $1 + 2\Delta$. The localized coupling loss deteriorates the soliton quality in spite of their compensation by the gain increase, according to (9.44).

These coupling losses must be decreased as much as possible or compensated by a localized optical amplifier. That problem can be most efficiently addressed by regenerator distance reduction. It has been shown that the coupling loss influence can be entirely compensated by reduction of regenerator distance length, even when the coupling loss coefficient is 2 dB.

As for the influence of pump-signal loss on the soliton-regeneration process, we must note that, in general, the pump signal has a higher loss rate than the soliton carrier. (The soliton carrier wavelength should be $\lambda_s \approx 1.55$ μm, while the pump-signal wavelength has to be smaller than λ_s, for example, at point $\lambda_p \simeq 1.45$ μm). The higher loss rate of the pump signal means that the distribution of the Raman gain is nonuniform along the transmission line length, which can also cause a nonadiabatic amplification.

It has been discovered [7] that introduction of localized coupling loss at the pump-injection points can compensate the nonadiabatic effect of the Raman pump signal. By changing combinations of the coupling loss and the pump loss rate, the most desirable combinations can be found. For proper compensation, a pump-loss rate equal to 1.2 dB/km should be combined with a coupling loss equal to 0.5 dB; a pump loss rate of 1.5 dB/km with a coupling loss of 1 dB; while a 1.8-dB/km loss rate requires a 1.5-dB coupling loss.

9.2.2.4 Influence of Random Gain on Soliton Regeneration

In the preceding analysis of the soliton regeneration process, it was assumed that the amplification coefficient is stable along the optical line length. But it is quite impossible to realize soliton amplifiers with equal amplification coefficients, so the value of the amplifier gains will have random deviation over the long-distance transmission line. The random gain may strongly influence the soliton regeneration process, so it is important to study this effect from the point of view of the soliton regime stability for a long-distance transmission. The random gain effects influence was analyzed in [11], so we will present the main results of that analysis.

Under normalization, the differential equation related to the soliton random gain regeneration is given as

$$j\frac{\partial q}{\partial \xi} + \frac{1}{2}\frac{\partial^2 q}{\partial \tau^2} + |q|^2 q = -j\Gamma q + j\beta\frac{\partial^3 q}{\partial \tau^3} + j\sum_{n=1}^{N} \alpha_n q(\tau, \xi_n - 0)\delta(\xi - \xi_n) \quad (9.45)$$

where ξ, τ, Γ, and β are the common denotations for the length coordinate, the time coordinate, the loss coefficient, and the third-order dispersion coefficient, respectively. All these parameters are normalized (see Section 7.2). The values $\xi_n = (n - 1)\Delta\xi$, $n = 1, 2, 3, \ldots, N$ define the amplifier positions, while the parameters α_n concern the amplification coefficients. The coefficients α_n are chosen to have an average value, $\overline{\alpha}$, such that it exactly compensates the loss influence over the regenerative length. Thus, we have

$$\overline{\alpha} = \exp(\Gamma\Delta\xi) - 1 \quad (9.46)$$

To consider the variation of the solitons' energy in the optical fiber under random gain influence, it is convenient to introduce the function $F(\xi)$ [11], which represents the total energy, $E(\xi)$, or

$$E(\xi) = \int_{-\infty}^{\infty} |q(\tau, \xi)|^2 d\tau = F^2(\xi)E_0 \quad (9.47)$$

The real function $F(\xi)$ is given by

$$F(\xi) = \left[\prod_{n=1}^{N}(1 + \alpha_n)\right]\exp(-\Gamma\xi), \quad \xi_n < \xi < \xi_{n+1} \quad (9.48)$$

During the amplification process, the newly generated soliton has the amplitude $(1 + 2\alpha_n)\eta_0$, where η_0 is the soliton initial amplitude. The difference in the energy between the amplified pulse and the stationary formed soliton is radiated out in a form of linear dispersive wave.

Since the function $F(\xi)$ is the solution of the linear differential equation

$$\frac{\partial F}{\delta\xi} = -\Gamma F + \sum_{n+1}^{N} \alpha_n F(\xi_n - 0)\delta(\xi - \xi_n) \quad (9.49)$$

the function $F^2(\xi)$ can be considered as the energy of the linear pulse in an ideal dispersionless medium. Equation (9.49) gives a Markov process,

$$F_n = F(\xi_n + 0) = (1 + \alpha_n)\exp(-\Gamma\Delta\xi)F(\xi_{n-1} + 0) = (1 + g_n)F_{n-1} \quad (9.50)$$

where

$$F_0 = \exp(\Gamma \Delta \xi) \tag{9.51}$$

and

$$g_n = (\alpha_n - \overline{\alpha}) \exp(-\Gamma \Delta \xi) \tag{9.52}$$

The probability function $P(F_n, \xi_n)$ of finding F_n at the point $\xi_n + 0$ is given as

$$P(F_n, \xi_n) = \int_0^\infty H(F_n | F_{n-1}, \Delta \xi) P(F_{n-1}, \xi_{n-1}) dF_{n-1} \tag{9.53}$$

where $H(F_n | F_{n-1}, \Delta \xi)$ is the conditional probability. Under the assumption that the average value and the variance of amplification coefficients α_n are equal to $\overline{\alpha}$ and $2D$, respectively, the following relations are valid:

$$\lim_{\Delta \xi \to 0} \frac{<F_n - F_{n-1}>}{\Delta \xi} = \lim_{\Delta \xi \to 0} \frac{<\alpha_{n+1} - \overline{\alpha}>}{\Delta \xi} F_n = 0 \tag{9.54}$$

$$\lim_{\Delta \xi \to 0} \frac{<(F_n - F_{n-1})^2>}{\Delta \xi} = \lim_{\Delta \xi \to 0} \frac{<(\alpha_{n+1} - \overline{\alpha})^2>}{\Delta \xi} F_n^2 = 2DF_n^2 \tag{9.55}$$

From (9.53), we can obtain the so-called Fokker-Planck equation for $P(F, \xi)$:

$$\frac{\partial P}{\partial \xi} = D \frac{\partial^2}{\partial F^2} (F^2 P) \tag{9.56}$$

The solution of (9.56) for the initial condition $P(F, 0) = \delta(F - F_0)$ is [11]

$$P(F, \xi) = \frac{1}{F\sqrt{4\pi D\xi}} \exp\left[-\frac{1}{4D\xi} \left(\log \frac{F}{F_0} + D\xi \right)^2 \right] \tag{9.57}$$

This solution indicates that the total energy flow is defined by

$$<E(\xi) - E_0> = [\exp(2D\xi) - 1)]E_0 \tag{9.58}$$

It is important now to estimate how much the stationary soliton energy differs from $E(\xi)$. At the point $\xi_n + 0$ right after the amplification for a small value of n, the soliton pulse is described by the function

$$q_{(S)}(\tau,\ \xi_n + 0) = \eta_n \operatorname{sech}[\eta_n(\tau - \vartheta_n)]\exp(j\sigma_n) \qquad (9.59)$$

where ϑ_n and σ_n define the center of the soliton and its phase. The soliton amplitude, η_n, is given by

$$\eta_n = \eta_{n-1}[1 + 2\alpha_n + \Theta(\alpha_n^2)]\exp(-2\Gamma\Delta\xi) \qquad (9.60)$$

The amplitude deviation in order α_n appears due to interactions between the soliton and the generated continuous wave. When this interaction is not strong, the soliton amplitude becomes

$$\eta_n = (1 + 2\alpha_n)\eta_{n-1} \qquad (9.61)$$

Equation (9.60) shows that the soliton energy after the nth amplifications can be written as

$$E_{n(s)} = \int |q_{(s)}|^2 d\tau = (1 + 2\alpha_n + \delta\alpha_n^2)\exp(-2\Gamma\Delta\xi)E_{n-1(s)} \qquad (9.62)$$

where the factor $(1 - \delta)$ is a measure of the energy that left the soliton and diffused to the dispersive wave. The ratio R_n of the soliton energy to the total energy after nth amplifications can be obtained from (9.48) and (9.62), that is,

$$R_n = \frac{E_{n(s)}}{E_n} = \frac{1 + 2\alpha_n + \delta\alpha_n^2}{(1 + \alpha_n)^2}R_{n-1} \qquad (9.63)$$

The following explanation, referring to the soliton dispersive wave interaction, can be given. If $\delta = 1$, all the energy is occupied by the newly formed stationary soliton. If $\delta = 0$, the interaction between the soliton and the dispersive wave does not occur, and the radiated dispersive wave is independent of the newly formed soliton. If $0 < \delta < 1$, the energy of the dispersive wave is being occupied by the soliton. And, finally, if $\delta > 0$, the energy of the soliton goes to the dispersive wave.

We can notice that the value $\overline{\alpha} \simeq \Gamma\Delta\xi$ is rather small. Hence, the deviation of the soliton energy from the total energy is small, as well. The variations of the total energy have a random character due to a random character of amplification coefficients α_n. The value δ can be used as a measure of the soliton transmission quality over the line with a large number of the random-gain amplifiers. Thus, the deviation of the stationary soliton energy from the total pulse energy can be estimated. The value of parameter δ can be estimated numerically for different values of lengths $\Delta\xi$ and gain variance $2D$, as in [11]. It has been shown that the proper choice of regeneration distance $\Delta\xi$ is essential for the stable soliton transmission over long-length optical-fiber lines. By the proper choice of a regenerative distance, unfavorable influence of the random gains will be eliminated.

The random character of the amplification coefficients will not bring into danger the stable transmission of the soliton pulses, but it will strongly influence the digital transmission quality. That is because the group velocity of an individual separate soliton pulse depends on the soliton amplitude, and this dependence will cause the time jitter of the soliton pulses and BER increase. This effect is analyzed in Section 9.3.

9.2.2.5 Soliton Regeneration by Lumped Erbium-Doped Optical Amplifiers

It was quite natural to assume that distributed amplification allowing for near cancellation of optical-fiber loss was necessary for long-distance optical soliton transmission, because the solitons involved continual balance between the dispersive and the nonlinear parts of the Schroedinger equation (6.59). It also was logical to start with Raman distributed amplifiers, as analyzed in Section 9.2.2.3. But with the rapid development of the more practical erbium-doped fiber optical amplifiers, attention has been changed to the possibility of regeneration of solitons by a chain of lumped erbium-doped fiber amplifiers. It was shown in [12] that such a transmission is quite efficient when the distance between amplifiers, L, is very short in comparison with soliton period, L_0, which is defined by (7.18), that is, for $L \ll L_0$.

If the soliton pulse width τ is compatible with the multigigabit-rate pulsewidths (typically 30 to 50 ps), the realm $L_0 \gg 50$ km corresponds to the dispersion D of, at most, a few picoseconds per nanometer kilometer in the wavelength region around 1,500 nm in dispersion-shifted single-mode optical fiber. This low-value dispersion is advantageous because it reduces the Gordon-Haus effect (see Section 7.4) and scales soliton-pair interaction to insignificance for spacing longer than 5τ [13].

Both the soliton energy fluctuations and the dispersion fluctuations (usually of about ± 0.5 ps/nm km for lengths up to 20 km) in dispersion-shifted optical fibers have a similar effect to long-distance soliton propagation. As energy and dispersion vary along the optical-fiber path, only their path average matters, as long as the average is over paths shorter in comparison with soliton period. Thus, by keeping the path-average power constant and equal to the usual soliton power from one period to the next, stable soliton transmission can be achieved [12]. To analyze the propagation of the solitons through the chain of erbium-doped lumped optical fiber amplifiers, following the analysis presented in [12], we should start from (6.60). This equation for the fiber with loss can be rewritten in the form

$$-j \frac{\partial u}{\partial z} = \frac{\delta(z)}{2} \frac{\partial^2 u}{\partial u^2} + |u|^2 u + j\Gamma u \qquad (9.64)$$

In (9.64), $\delta(z)$ is a normalized group velocity dispersion parameter defined as

$$\delta(z) = D(z)/D_{av} \tag{9.65}$$

where $D(z)$ is the local dispersion parameter, defined in (6.60) and (J.4), and D_{av} is the path-averaged dispersion.

If the condition $L \ll L_0$ is satisfied, both dispersion and nonlinear effects can be treated as perturbations. Taking only the first term in the perturbation expansion, (9.64) becomes

$$\frac{\partial u}{\partial z} = -\Gamma u \tag{9.66}$$

The lowest-order solution of (9.66) takes the form

$$u(z, t) = u_0(t)\exp(-\Gamma z) \tag{9.67}$$

In (9.67), u_0 is not a function of distance z, but the solution of (9.65) can significantly depend on the distance z. Hence, the function $u_0(t, z)\exp(-\Gamma z)$ can be introduced as a solution of (9.65). Equation (9.65) now becomes

$$-j\frac{\partial u_0}{\partial z} = \frac{\delta(z)}{2}\frac{\partial^2 u_0}{\partial u^2} + |u_0|^2 u_0 e^{-\gamma z} \tag{9.68}$$

where $\gamma = 2\Gamma$ presents the optical-energy loss parameter. This equation can be integrated over the length interval $\delta z = L$ and after that normalized by distance L, so we have

$$-j\frac{\Delta u_0}{\Delta z} = \frac{1}{L}\int_{nL}^{(n+1)L}\frac{\delta(z)}{2}\frac{\partial^2 u_0}{\partial u^2}dz + \frac{1}{L}\int_{nL}^{(n+1)L}|u_0|^2 u_0 e^{-\gamma z}dz \tag{9.69}$$

The terms with function u_0 can be placed outside the integrals from (9.69) since they do not vary significantly over distance L. The second integral in (9.69) is a constant, because of the periodical character of the amplification process. This constant, which is equal to

$$\chi = \frac{1 - \exp(-\gamma L)}{\gamma L} \tag{9.70}$$

presents the ratio of the path-averaged power to the peak power at the distance L. Besides, the first integral, which represents dispersion influence, can be replaced by the averaged value $\delta_{av}(z)$ in the interval $nL < z < (n + 1)L$, or

$$\delta_{av}(z) = \frac{1}{L} \int_{nL}^{(n+1)L} \delta(z)dz \qquad (9.71)$$

Now, by treating Δz as a differential, (9.69) takes the form of a nonlinear Schroedinger equation, that is,

$$-j\frac{\partial u_0}{\partial z} = \frac{\delta_{av}(z)}{2}\frac{\partial^2 u_0}{\partial u^2} + \chi\,|u_0|^2 u_0 \qquad (9.72)$$

The averaged value $\delta_{av}(z)$ approaches unity value if dispersion D, averaged over distance L, approaches the value D_{av}. According to (9.72), the optical power of each soliton pulse at the beginning of each amplification cycle is

$$P_n = P_o/\chi \qquad (9.73)$$

where P_o represents the soliton pulse power at the optical-fiber input or the soliton pulse power in a lossless optical fiber. Now we can conclude that the path-averaged power is equal to P_o, which is one of the most important conclusions.

The authors of [12] concluded that variation in dispersion D need not be as restrictive as it was in the previous analysis. To extend the distance for achieving the prescribed value D_{av}, the integrals in (9.69) can extend over not just one but m amplification periods, as long as $mL \ll L_0$. Practically speaking, that means acceptably stable soliton transmission will be observed as long as the only significantly large Fourier components of $D(z)$ are those of wavelength considerably less than L_0. By choosing the optical fibers with such dispersion parameters at 1,560-nm wavelength that L_0 is around 1,000 km, the total nonlinear transmission system length, L_Σ, can reach the value $L_\Sigma = 8L_0$ for multigigabit rates (corresponding to the soliton pulse width of 30 to 50 ps). The distance between two subsequent optical amplifiers in that case can be up to 100 km [13], although such long amplifier spacing usually is not desirable because of the extent of spontaneous emission noise in erbium-doped fiber amplifiers (see Section 8.4.1).

9.3 EVALUATION OF BER IN THE NONLINEAR SYSTEM

We will discuss two possible schemes of nonlinear lightwave systems. In the first, transmission is made in a single optical channel by a single carrier wavelength. In the second, transmission is realized by employment of the WDM technique; thus, multichannel transmission is realized. Time jitter due to different group velocities of separate soliton pulses is the cause of a bit error in a single-transmission channel. The amplitudes and carrier wavelengths of individual pulses in a soliton pulse train are not identical due to the source and regenerator instability, having random variations around some average values. These variations will lead to the difference in group velocities between individual soliton pulses in a soliton train, causing time jitter at the receiver end and increased BER.

The BER in a multichannel nonlinear transmission system is determined mainly by the signal-to-interference ratio. The interference signal appears due to cross-talk between individual wavelength channels. The cross-talk is caused mainly by stimulated scattering in a single-mode optical fiber. Since the scattered signal has a different wavelength from that of the pump signal, there is the possibility that the scattered-signal wavelength is the same as the carrier wavelength of another channel.

We will analyze the transmission quality in a single-channel nonlinear system first and then consider multichannel nonlinear transmission.

9.3.1 Error Probability in a Single-Channel Nonlinear System

Considering the main statements expressed in this chapter, we can emphasize the most important characteristics of the solitons during their propagation through an optical fiber, which are as follows:

- The group velocity of the soliton pulse depends on the pulse amplitude, which means that a pulse with high amplitude is faster than a pulse with small amplitude.
- The pulsewidth of the soliton pulse is inversely proportional to its amplitude at any moment.
- The pulsewidth depends primarily on the fiber loss, which means that intersymbol interference during soliton propagation through the optical fiber can be expected.
- The group velocity of the soliton pulse is also a function of wavelength (by the refractive index and its derivatives), so the random deviations of the soliton carrier wavelength will cause fluctuations in the pulse group velocity.

These characteristics of soliton pulses are illustrated in Figure 9.3. It can be seen that the spacing between two adjacent soliton pulses changes with time, which leads to time jitter at the optical receiver. At the same time, it is clear that the spacing can decrease below a critical value, so that mutual interaction can appear. It is reasonable to assume that the initial separation between two soliton pulses is sufficient to prevent mutual interaction at a regenerative distance (when $z < z_1$). Consequently, the duty cycle of the soliton-pulse waveform must be lower than a limit value. In [7] it was recommended that the duty cycle should be less than 15%. Nevertheless, when $z > z_1$, the changes in the pulse separations might result in mutual interaction. Thus, when the duty cycle almost reaches the limit value ($\approx 15\%$), the soliton pulses must be reshaped. That means that a repeater is required at approximately every $z = z_1$ km. The basic role of the repeater is only to amplify the soliton amplitude, but this

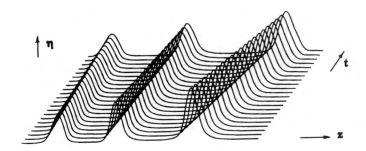

Figure 9.3 The change in the initial parameters of a soliton train during propagation through an optical fiber. (*Source:* [14], © Chapman and Hall, reprinted with permission.)

process is accompanied by the pulse reshaping. Hence, BER is evaluated for the regenerative nonlinear system model.

To evaluate the BER at the output end of the optical-fiber line, caused by time jitter of the solitons, an appropriate model that will describe the relative fluctuations of the soliton pulses around the initially defined positions must be chosen. Since the group-velocity fluctuations of the soliton pulses are directly related to fluctuations of the pulse amplitude and the soliton-carrier wavelength, the fluctuations of amplitude η and wavelength λ should be described first. Then, using the functional relation

$$v_g = f(\eta, \lambda) \tag{9.74}$$

the fluctuations of group-velocity, v_g, should be described.

The pulse-amplitude fluctuations have a random character and can be characterized by the probability distribution function $p(\eta)$. Since the soliton amplitude, η, originates from an initial amplitude, η_0, it is necessary to know the distribution function $p_0(\eta_0)$ of the initial amplitudes and then to find the function $p(\eta)$. We must notice that

$$\eta = f_1(\eta_0, A, \Gamma, z_1) \tag{9.75}$$

where A is the random gain in a line regenerator, Γ is the normalized loss coefficient, and z_1 is the length of the regenerative distance. Since Γ and z_1 are not random variables, it is clear that the problem of finding an amplitude distribution function will be reduced to transformation of probabilities $p_0(\eta_0)$ and $p_A(A)$ to the function $p(\eta)$, according to the functional dependence in (9.75).

Similar considerations can be applied to the carrier-wavelength fluctuations of individual pulses in an information train. It is necessary to know the distribution function $p_1(\lambda)$ for the employed light source and then to find the function $p_2(\lambda)$ that reflects the carrier-wavelength fluctuations during the amplification process. The function $p_2(\lambda)$ is dependent on the following parameters: $p_1(\lambda)$, η, and the noise (continuous wave) level. It is clear that the distribution function $p_v(v_g)$ will be dependent on functions $p(\eta)$ and $p_2(\lambda)$, so the following general relation is valid:

$$p_v(v_g) = p_v[f(\eta, \lambda), p(\eta), p_2(\lambda)] \tag{9.76}$$

The BER can be evaluated after the evaluation of the group-velocity distribution function by averaging over the assemblage of group velocities. But first it is necessary to answer the question of what referent positions of the individual soliton pulses should be chosen to evaluate the time jitter of the pulses with group velocity v_g in an information train. The appropriate measure for a time jitter is the group delay of the soliton pulse, defined as $t_g = Nz_1/v_g$, where N is the total number of regenerators on the optical-fiber line. Since the referent position of an individual pulse corresponds to the decision point at the optical receiver, it is reasonable to assume that the referent time positions of the pulse in the pulse train correspond to the maximum widths of the eye-pattern diagram. The relevant eye-pattern diagram is that referring to the pulse stream that propagates with an average velocity $\bar{v}_g$. Now the decision point t_0 at the optical receiver is defined as

$$t_0 = Nz_1/\bar{v}_g \tag{9.77}$$

Under the assumption that the soliton pulses, reaching the optical receiver end, have the form

$$q(z, t) = \eta \operatorname{sech}\left(\frac{t - Nz_1/v_g}{T}\right) \tag{9.78}$$

where the value $2T$ is the soliton pulsewidth, the function of the detected electrical pulse in the photodetector can be found. The detected electrical pulse has the following value at the decision moment t_0:

$$s(t_0, z) = S \operatorname{sech}\left(\frac{t_0 - Nz_1/v_g}{T}\right) = S \operatorname{sech}\left(\frac{(v_g - \bar{v}_g)z_1 N}{v_g \bar{v}_g T}\right) \tag{9.79}$$

where $S = 0.5 \Re c \eta_0^2 S_c n \epsilon_0$ ($\Re$ is the responsivity of the photodetector, c is the light velocity, S_c is the cross-sectional area of the fiber core, and ϵ_0 is the dielectric permittivity in the vacuum). Equation (9.79) can also be considered to be a dependence, $s(v_g)$.

The BER evaluation in a nonlinear lightwave system can be done starting from the expression for BER in a linear lightwave system [see (E.13)], which corresponds to the case when jitter is absent, that is,

$$P_0 = 0.5 \ \mathrm{erfc}\{[s(\bar{v}_g) - b]/\sqrt{2}\sigma_n\} \tag{9.80}$$

In (9.80), b is the decision threshold, and σ_n is the standard deviation of the total noise in the system. In the evaluation process, the parameters b and σ_n should be chosen in such a way that the error probability, P_0, is equal to a prescribed value. For example, if the prescribed value is $P_0 = 10^{-9}$, it is necessary that $s(v_g) - b \simeq 6\sigma_n$.

The influence of the time jitter on the value of P_0 can be determined by averaging the instantaneous bit rate over the assemblage of group velocities. Thus, the expression for the BER becomes

$$P = \int_0^\infty 0.5 p_v(v_g) \mathrm{erfc}\left(\frac{s(v_g) - b}{\sqrt{2}\ \sigma_n}\right) dv_g \tag{9.81}$$

The described procedure for BER evaluation in a nonlinear lightwave system is rather complex, and the application of numeric methods is necessary. To obtain the result as precisely as possible, it is necessary to choose the proper distribution functions for instantaneous initial pulse amplitudes, instantaneous pulse wavelengths, and amplifier coefficients assemblage. The simplest way is to approximate these functions by Gaussian curves, but more realistic functions, such as those in [11], can be used.

By that described method, the BER in a nonlinear lightwave system was evaluated in [14], but only the influence of the pulse-amplitude variation was taken into account. Obtained results are very illustrative, since they predict the limit performances of a nonlinear transmission system. It was found that the time jitter induced by the group-velocity variations of soliton pulses increases the BER in a nonlinear transmission system and limits the total line length. (The line length is the transmission distance without optoelectronic conversion of the pulses.) The main conclusion is that the upper bit-rate bound of practically acceptable nonlinear transmission is about 50 Gb/s. Below this bit rate, the advantage of nonlinear signal transmission over the best case of linear transmission is evident, and considerably longer line lengths can be achieved.

On the other hand, Gordon and Haus [15] evaluated the influence of the wavelength fluctuation–influenced jitter. The random shift in the soliton carrier frequency (or wavelength) during the soliton amplification process is calculated to be [see (7.88)]

$$\Delta\omega = \frac{2\pi c(\lambda_1 - \lambda_2)}{\lambda_1\lambda_2} = \frac{2}{3}a\eta \tag{9.82}$$

where a reflects the nonsoliton part (noise) amplitude. Using this result, we can easily find the relation between carrier wavelengths before amplification (λ_1) and after amplification (λ_2):

$$\lambda_2 = \frac{2\pi c \lambda_1}{2\pi c + \frac{2}{3}a\eta\,\lambda_1}$$

(9.83)

In (9.83), η represents the stochastic amplitude described by (9.75).

It was estimated in [15] that the jitter influenced by carrier-wavelength fluctuation limits the transmission capacity of nonlinear systems to 23,600 GHz-km. In such a way, based on results presented in [14] and [15], we can predict the capacity of nonlinear lightwave systems to be within the range of 20,000 to 25,000 GHz-km.

9.3.2 SNR in a Nonlinear Multichannel System

Time jitter is the main factor causing bit error in a single-channel nonlinear lightwave system, but another factor that is important in a multichannel system is the cross-talked signal level. The cross-talked signal, or interference signal, can appear as a consequence of several nonlinear effects [16]. We will analyze cross-talk signal in Raman scattering, since that is widely regarded as the most severe limitation to transmission capacity in multichannel lightwave systems. The influence of other nonlinear effects will be mentioned after this analysis.

The nonlinear cross-talk effect due to stimulated Raman scattering can appear in both unidirectional and bidirectional transmission, since WDM can be used for both unidirectional and bidirectional transmission through the same optical fiber. These situations are illustrated in Figure 9.4. In the figure, two-channel transmission is illustrated, so the wavelengths λ_1 and λ_2 ($\lambda_1 > \lambda_2$) correspond to the first and the

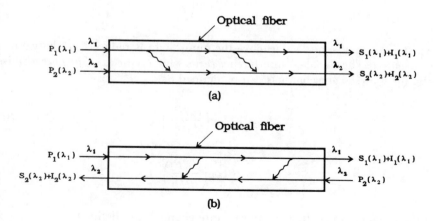

Figure 9.4 The model of crosstalk due to Raman scattering: (a) unidirectional, and (b) bidirectional transmission. (*After:* [17], Chapman and Hall, reprinted with permission.)

second channel, respectively. At the same time, it is useful to assume that the wavelength difference $\Delta\lambda = \lambda_1 - \lambda_2$ corresponds to the Raman scattering shift (see Chapter 6). The first-channel signal is transmitted by wavelength λ_1, while the second-channel signal is transmitted by wavelength λ_2.

Because in the stimulated Raman scattering process, the information signal leaves channel 1 and goes to channel 2, it is clear that the signal cross-talking might cause an increase of the BER in both channel 1 and channel 2. That is because of the signal decrease in channel 1 and the interference (i.e., noise) increase in channel 2. In an analysis of the influence of stimulated Raman scattering on the transmission quality in multichannel nonlinear system, only Stokes line influence is considered, while anti-Stokes line influence is neglected since it has a relatively small power level.

According to the model from Figure 9.4, the optical power of information optical signals is denoted by $P_1(\lambda_1)$ and $P_2(\lambda_2)$, the corresponding receiver power by $S_1(\lambda_1)$ and $S_2(\lambda_2)$, while the interference signals due to the appearance of stimulated Raman scattering are designated by $I_1(\lambda_1)$ and $I_2(\lambda_2)$. The wavelength λ_1 is considered to be the pump signal. If the cross-talking due to stimulated Raman scattering is defined as an attenuation, then the far-end cross-talk attenuation is relevant for unidirectional transmission, and the near-end cross-talk attenuation is relevant for bidirectional transmission. The far-end and near-end cross-talk attenuations are defined by the coefficients A_1 and A_2, concerning channels 1 and 2, respectively; that is,

$$A_1 = -10\ell og\left(\frac{|I_1|}{P_2}\right) \tag{9.84}$$

$$A_2 = -10\ell og\left(\frac{I_2}{P_1}\right) \tag{9.85}$$

I_1 is written as an absolute value since the pump light is reduced, so the value of I_1 will be negative. The signal-to-interference ratio, which is the measure of transmission quality, for the relevant channels can be defined as

$$R_1 = 10\ell og\left(\frac{S_1/P_1}{|I_1|/P_2}\right) \tag{9.86}$$

$$R_2 = 10\ell og\left(\frac{S_2/P_2}{I_2/P_1}\right) \tag{9.87}$$

Since the attenuation coefficients in separate channels are defined as

$$\gamma_i = 10\ell og(P_i/S_i), \; i = 1, \; 2 \tag{9.88}$$

the signal to interference ratios can be expressed in the form

$$R_i = A_i - \gamma_i \tag{9.89}$$

It can be seen that the expressions for unidirectional and bidirectional transmissions are formally identical, but it does not mean that the evaluated values will be identical for the same input parameters. That is because of the different character of forward and backward stimulated Raman scattering.

To evaluate the parameters R_i, it is necessary to consider the variations of the signal powers in the first and second channels, taking the standard values of the parameters that characterize optical fiber and the Raman process (see Chapter 6). It is convenient to make a separate analysis of unidirectional and bidirectional tansmission. It is also convenient to preserve the same denotation for pump and scattered signal as those in Chapter 6. So the signal power in channel 1, carried by wavelength λ_1, is denoted by P_p (pump power), and the signal power in channel 2, carried by wavelength λ_2, is denoted by P_r (Raman signal).

9.3.2.1 Signal-to-Interference Ratio for Unidirectional Transmission

Applying the Raman-scattering theory from Chapter 6, the variations of the powers of scattered and pump signals can be described by the following equations [17]:

$$\frac{dP_r}{dz} = \frac{\alpha}{S}P_p(z)P_r(z) - \gamma P_r(z) \tag{9.90a}$$

$$\frac{dP_p}{dz} = -\frac{\alpha}{S}P_p(z)P_r(z) - \gamma P_p(z) \tag{9.90b}$$

where S is the cross-sectional area of the fiber core, α is the Raman amplification coefficient, and γ is the loss coefficient in the optical fiber. The first term on the right side of (9.90a) and (9.90b) is related to an increase of scattered signal, as in (9.90a), or to a decrease of pump signal, as in (9.90b).

To solve (9.90a), we can assume that the pump signal decreases as

$$P_p(z) = P_{p0} \exp(-\gamma z) \tag{9.91}$$

where only the influence of the fiber loss is taken into account, while the influence of the Raman scattering is neglected because it is considerably smaller. By replacing the value of $P_p(z)$ from (9.91) in (9.90a), the following solution is obtained:

$$P_r(z) = P_{r0} \exp\left\{-\gamma z + \frac{\alpha P_{p0}}{S\gamma}[1 - \exp(-\gamma z)]\right\} \tag{9.92}$$

By setting the exponent to be zero, the threshold pump power, P'_{p0}, which determines the exponential increase of scattered power, can be obtained:

$$P'_{p0} = \frac{\gamma^2 Sz}{\alpha[1 - \exp(-\gamma z)]} \tag{9.93}$$

Now, by using the definition, (9.93), and the solution, (9.92), the far-end cross-talking attenuation can be found:

$$A_2 = -10 \; \ell og \left[\frac{P_r(L)|_{P_{p0}>0} - P_r(L)|_{P_{p0}=0}}{P_{p0}} \right] \tag{9.94}$$

$$= -10 \; \ell og \left(\frac{P_{r0} \exp(-\gamma L) \left\{ \exp\left[\dfrac{\alpha P_{p0}}{S\gamma} (1 - \exp(-\gamma L)) \right] - 1 \right\}}{P_{p0}} \right)$$

In the case of symmetrical multichannel transmission, when the initial powers are equal, or $P_{r0} = P_{p0} = P_0$, (9.94) becomes

$$A_2 = \frac{10\gamma L}{\ell n\, 10} - 10 \; \ell og\left(\exp\left\{ \frac{\alpha P_0}{S\gamma} [1 - \exp(-\gamma L)] \right\} - 1 \right) \tag{9.95}$$

Equation (9.95) can be shown as the function $A_2 = f(L)$, as in Figure 9.3, where P_0 is a parameter. The curve from Figure 9.3 has an absolute minimum in the point defined by $dA_2/dL = 0$, that is, by

$$\frac{S\gamma}{\alpha P_0} = \frac{\exp\left\{ \dfrac{\alpha P_0}{S\gamma} [1 - \exp(-\gamma L)] - \gamma L \right\}}{\exp\left\{ \dfrac{\alpha P_0}{S\gamma} [1 - \exp(-\gamma L)] \right\} - 1} \tag{9.96}$$

The point of the absolute minimum cannot be determined in a closed form, but an approximate value can be found. For example, if $\alpha P_0/(S\gamma) < 0.1$, corresponding to $P_0 \simeq 15$ mW, the point of the minimum is

$$L_{min} = \frac{\ell n\, 2}{\gamma} \tag{9.97}$$

while the function minimum is

$$A_{2min} = 10 \, \ell og \frac{\alpha P_0}{4S\gamma} \tag{9.98}$$

Since it is expected in all real situations that

$$\frac{\alpha P_0}{S\gamma} < 0.1 \tag{9.99a}$$

and

$$\exp(-\gamma L) < 0.1 \tag{9.99b}$$

(9.95) can be simplified and takes the form

$$A_2 = \frac{10\gamma L}{\ell n \, 10} - 10 \, \ell og\left(\frac{\alpha P_0}{S\gamma}\right) \tag{9.100}$$

The signal-to-interference ratio, according to (9.88), (9.89), and (9.95), now becomes

$$R_2 = -10 \, \ell og\left\{\frac{\alpha P_0}{S\gamma}[1 - \exp(-\gamma L)] - 1\right\} \tag{9.101}$$

For long lengths, (9.101) tends toward the limit value

$$R_{2\infty} = -10 \, \ell og\left[\exp\left(\frac{\alpha P_0}{S\gamma}\right) - 1\right] \simeq -10 \, \ell og\left(\frac{\alpha P_0}{S\gamma}\right) \tag{9.102}$$

For example, for the typical values of parameters α, γ, and S ($\alpha = 3 \cdot 10^{-11}$ cm/W, $y = 0.4$ dB/km, $S = 50 \ \mu m^2$) and for $P_0 \approx 1$ mW, the parameter $R_{2\infty}$ is approximately 20 dB.

The reduction of the pump power can be evaluated from (9.90b) under the same assumptions as those already mentioned. Thus, it is obtained that the far-end signal attenuation A_1 is

$$A_1 = \frac{10\gamma L}{\ell n \, 10} - 10 \, \ell og\left\{1 - \exp\left[-\frac{\alpha P_{p0}}{S\gamma}(1 - \exp(-\gamma L))\right]\right\} \tag{9.103}$$

This value of parameter A_1 is almost identical with the value A_2 in the region of optical powers that are of interest (for $P_0 \leq 10$ mW). The parameter R_2 for channel 1 is

$$R_1 = -10 \, \ell og \left\{ 1 - \exp\left[-\frac{\alpha P_{p0}}{S\gamma}(1 - \exp(-\gamma L)) \right] \right\} \qquad (9.104)$$

and, therefore, it is approaching the value of R_2 for the condition (9.99a).

Obtained results are graphically illustrated in Figure 9.5 for two channels with digital return-to-zero signals. Whenever there is a mark in both channels, the signal power in channel 1 is decreased, and the signal power in channel 2 is increased due to stimulated Raman scattering. Whenever there is a space in either channel, there are no changes in bit pattern. (It must be noted that in conventional cross-talk a mark in channel 1 can always produce a signal in channel 2.) Thus, it is clear that the influence of the stimulated scattering effect is not symmetric for both channels. The widths of eye-pattern diagram in channel 1 will have a partial decrease, while the eye-pattern diagram in channel 2 will have a partial improvement. So the overall effect of stimulated Raman scattering is a degradation in SNR for the short-wavelength channel and no degradation for the longer-wavelength channel.

9.3.2.2 Signal-to-Interference Ratio for Bidirectional Transmission

In the case of bidirectional transmission through the same optical-fiber, the signal in channel 2 propagates in the negative direction of the z coordinate. Due to this fact, the differential length, dz, in (9.90a) must be replaced by $-dz$. By solving such

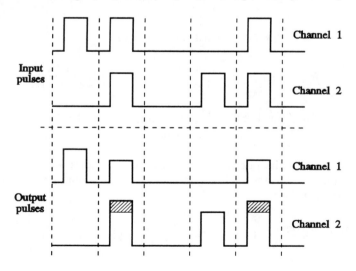

Figure 9.5 The illustration of channels crosstalk due to stimulated Raman scattering. The shadowed areas denote increase or decrease of initial bit powers.

a changed equation, with boundary condition (9.91), the power $P_r(z)$ can be expressed as

$$P_r(z) = P_{r0} \exp\left\{-\gamma(z - L) + \frac{\alpha P_{p0}}{S\gamma}[\exp(-\gamma z) - \exp(-\gamma L)]\right\} \qquad (9.105)$$

The attenuation A_2 can be now determined by

$$A_2 = -10 \, \ell og\left[\frac{P_r(0)|_{P_{p0}>0} - P_r(0)|_{P_{p0}=0}}{P_{p0}}\right] \qquad (9.106)$$

To find the attenuation coefficient A_1, we must solve (9.90b) for these initial conditions:

$$P_p(0) = P_{p0} \qquad (9.107a)$$

and

$$P_r(z) = P_{r0} \exp[-\gamma(L - z)] \qquad (9.107b)$$

It can be easily shown that the values of the attenuation coefficients A_1 and A_2, as well as the values of the signal-to-interference ratios R_1 and R_2, in a bidirectional transmission system are equal to the same values in a unidirectional transmission system. That means the same conclusions are valid for both unidirectional and bidirectional transmission in a multichannel optical system.

The analyses here and in Section 9.3.2.1 were related to a nonlinear system with applied intensity modulation. The question now is what will happen when the angular modulation is applied. Such a case corresponds to the constant amplitude of optical signals in separate channels, so the gain in channel 2 does not depend on the bit pattern and tends to be not sporadic but uniform. We can expect that the decrease in optical power in channel 1 will be considerably smaller. Hence, the angular modulation of optical signals will cause a reduction of the interference due to stimulated Raman scattering. The reduction is higher for the larger difference in the group velocities of separate channels [17]. Since the group velocity is a function of the modulation frequency of the optical signal, the larger difference between modulation frequencies of separate channels will cause reduction of the interference among the channels.

9.3.3 Some Design and Performance Issues of Multichannel Systems

Two optical spectral regions (or optical frequency bands) in a low-loss optical fiber are available for multichannel lightwave system design. These regions are illustrated in Figure 9.6. The carrier optical wavelength of the optical channels should be separated enough to prevent mutual overlapping of neighboring-channel spectra. Wavelength-tunable optical lasers are used to generate optical carrier wavelengths (see Section 3.3.5). Different types of optical multiplexers can be used to combine separate optical channels [18], while the multiple channels can be demultiplexed at the receiver by the use of frequency-selective components such as optical gratings or bandpass filters.

Either an FDM technique with a coherent detection scheme (see Section 5.3) or a WDM technique in combination with optical filters can be used for multichannel optical transmission. That means the received optical signal can be demultiplexed into individual channels by the use of electrical or optical techniques. An FDM technique has the potential for exploiting the wide bandwidth from Figure 9.6, and even several thousands of channels can be transmitted if the channel spacing is reduced to a few GHz and homodyne coherent detection is applied (see Chapter 5). Hence, we can say that an FDM technique implicitly means that the channel spacing is lower than 10 GHz. If the channel spacing is larger and above 1 nm, such a system is referred to as a WDM system.

The maximum number of optical channels that can be packed in the wavelength regions in Figure 9.6 is determined by the level of the inteference from the other channels. The interference-signal level was evaluated in Chapter 5 for FDM coherent detection lightwave systems, without taking into account the influence of nonlinear effects. But, as was shown in Chapter 6 and Section 9.3.2, several nonlinear effects can lead to interchannel cross-talk. That means both the intensity and the phase of one channel are influenced by the optical signals from neighboring channels.

In this section, we will estimate the number of channels that can be packed into the wavelength regions in Figure 9.6. Our primary interest is the wavelength

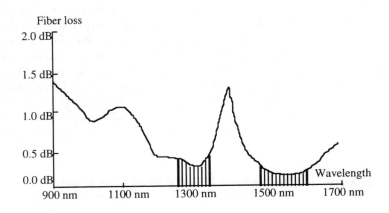

Figure 9.6 The spectral regions in a low loss optical fiber used for multichannel transmission.

region around 1,550 nm, because it is wider and the optical fiber losses are lower. If we assume that channel spacing is y times the bit rate R_d (y is about 4 for heterodyne detection schemes; see Section 5.3.2), the optical frequency width of one optical channel can be defined as

$$\Delta f_s = y R_d \qquad (9.108)$$

If the optical frequency range accessible to a tunable semiconductor laser or receiver filtering scheme is Δf_Σ, the following relation can be written:

$$\Delta f_\Sigma = M \Delta f_s \qquad (9.109)$$

where M represents the number of optical channels. The total data capacity of a multichannel lightwave system can be expressed through the product MR_d, or

$$MR_d = \Delta f_\Sigma / y \qquad (9.110)$$

Generally speaking, the parameter Δf_Σ is a sensitive function of the tuning speed from one carrier wavelength to another [19]. To spread the value Δf_Σ, more than one tunable laser or optical receiver (i.e., filter) can be used, if necessary. The current state of the art for the tuning range of the semiconductor laser is up to 100 nm [20], but we can expect further increase of this value.

Having in mind that analyses from Chapter 6 and this chapter, we can say that nonlinear effects in an optical fiber are likely to impose severe restrictions on transmitter power, especially in densely packed FDM coherent systems. Stimulated Brillouin scattering and induced cross-talk will be of concern only in bidirectional multichannel lightwave systems, even if optical powers injected by lasers into single-mode optical fibers do not reach the critical level. However, this type of cross-talk is easily avoided with proper design of the multichannel lightwave system. The cross-talk can play an important role only if the channel spacing matches almost exactly the Brillouin frequency shift, about 10 to 11 GHz for the 1,550-nm wavelength region (see Chapter 6). Such an exact matching can be easily avoided.

The self-phase modulation effect, explained in Section 6.2.3, can be an important source of the interference noise in multichannel lightwave systems. In this case, the term *self-phase* should be replaced by *cross-phase*, because the intensity-dependent phase shift for a specific channel depends not only on the power of that channel but on the power of the other channels, as well. The total phase shift presents the sum of the phase shifts caused by other individual channels, and that means that situation is the worst for coherent multichannel systems, not only because of the large number of channels, but because of the phase-sensitive nature of the optical receiver (see Section 5.3.4).

The four-wave mixing effect is important whenever channel spacing is small enough and optical fiber dispersion satisfy the phase-matching condition. It is quite clear if we keep in mind that a four-mixing wave can arise whenever three waves propagate through the same fiber. In an M-channel lightwave system, there can be a number of such three-wave combinations. In the case of equal spacing between neighboring channels, the new generating frequencies coincide with the existing frequencies, which causes the intensive power transfer between various channels.

Raman cross-talk has been analyzed in detail in this chapter, and now we can only point out that it can be avoided if channel powers are so small that Raman amplification is negligible over the optical-fiber length. The level of the interference noise caused by Raman cross-talk increases rapidly if the total frequency range, Δf_Σ, of all channels in the multichannel lightwave system becomes comparable or higher than the Raman frequency shift illustrated in Figure 6.2.

A comparison of the influence of various nonlinear effects on multichannel lightwave system characteristics has been done in several papers [16, 21, 22]. It was concluded that the four-wave mixing effect has the predominant influence for the number of channels up to 20. If the number of channels is higher than that value, the cross-phase modulation effect becomes dominant. The Raman cross-talk effect imposes the most severe restrictions on transmitter power if the number of channels becomes higher than several hundred.

Several methods have been proposed so far to compensate for the influence of the nonlinear effects on lightwave systems characteristics. This is important not only for multichannel FDM and WDM lightwave systems, but for long-length transmission systems, as well. The four-wave mixing effect can be suppressed by using two incoherent polarized light sources in multichannel transmission [23] or by using optical multiplexers/demultiplexers and delay line [24]. The effect of stimulated Brillouin scattering can be suppressed by applying sinusoidal strain distribution to the optical fiber [25] or by employing externally modulated lasers, such as in [22].

9.3.3.1 Optical Filters for Multichannel Lightwave Systems

Optical filters were mentioned several times in this book. They play a very important role not only in channel selection in a multichannel lightwave system, but also as a complement to the optical amplifiers. Because of that, we will consider the main characteristics of the optical filters that are used in lightwave transmission systems. A tunable optical filter placed just before the optical receiver is needed in a multichannel WDM system, while this filter should accompany the optical amplifier in various other situations.

The bandwidth of the optical filter should be large enough to be a window for the desired channel, but small enough to block the noise due to cross-talk or amplified spontaneous emission. A Fabry-Perot interferometer is commonly used as a physical case for tunable optical-filter design, but some other effects such as acousto-optic or electro-optic effects applied in the form of a compact $LiNbO_3$ waveguide device can be used as well [26, 27]. The optical filter function can be based not only on the restrictive role but on an active supportive role. In that case, the tunable optical filters operate on the principle of amplification of the selected frequency band of incoming optical signal, making a difference in level between the signal from the selected band and the signal from outside the band. Basically, these optical filters play a role of optical amplifiers applied in a different function.

The physical mechanism of the Fabry-Perot interferometer–based optical filter was analyzed in Section 3.2.4.3. The interferometer length is adjusted electronically by piezoelectric control. The wavelength characteristics of the Fabry-Perot filter are wavelength dependent, because the filter transmits only those optical frequencies that correspond to the longitudinal frequencies of the resonator. The frequency spacing between the two neighboring longitudinal mode frequencies is given as

$$\Delta f = \frac{c}{2nL} \tag{9.111}$$

where n is the refractive index of the material in the resonator, L is resonator length, and c is the light velocity in free space. The 3-dB bandwidth Δf_c of the filter is related to Δf by the parameter F, which presents filter finesse defined as

$$\Delta f_c = \Delta f / F \tag{9.112}$$

Optical filter parameters are illustrated in Figure 9.7.

Finesse F is a function of resonator losses. For an ideal resonator with mirrors of reflectivity r and transmission $t = 1 - r$, finesse is given as [19]

$$F = \pi \sqrt{r/t} \tag{9.113}$$

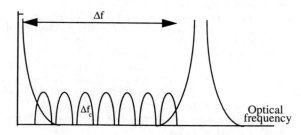

Figure 9.7 Tunable optical filter parameters.

Finesse of Fabry-Perot tunable optical filters are in the range 150–300. A transmission peak of the filter can be tuned through one spectral region Δf by piezoelectrically tuning length L through one optical half-wavelength. The tuning range depends on the filter length, according to (9.99). For a tuning range of about 100 GHz, the filter length should be $L < 1$ mm for a refractive index value of $n = 1.5$. For a wider range of tunability, the filter length should be as small as several micrometers. The number of channels, M, that can fit in the spectral region Δf is given approximately by [19]

$$M = \Delta f / \Delta f_s \cong F/6 \qquad (9.114)$$

where Δf_s is defined by (9.96). The number of channels can be easily calculated from (9.12). If the channel reflectivity is $r = 0.99$, the number of channels is more than 100.

In conclusion, we can point out the numerous experimental verifications of the theoretical results presented in this chapter. We mentioned only a few [28–30] that can be the great stimulus for future efforts in the practical realization of nonlinear lightwave systems. The next generation of telecommunications engineers will witness various achievements not only in the domain of nonlinear transmission systems but of combined coherent nonlinear systems, as well.

REFERENCES

[1] Mitsche, F. M., and L. F. Mollenauer, "Stabilizing the Soliton Laser," *IEEE J. Quantum Electron.*, QE-22(1986), pp. 2242–2250.

[2] Mollenauer, L. F., and R. H. Stolen, "The Soliton Laser," *Optics Lett.*, 9(1984), pp. 13–15.

[3] Nakazava, M., et al., "Transform Limited Pulse Generation in Gigahertz Region From Gain Switched Distributed Feedback Laser Diode Using Spectral Windowing," *Optics Lett.*, 15(1990), pp. 715–717.

[4] Suzuki, M., et al., "Transform Limited 14 ps Optical Pulse Generation With 15 GHz Repetition Rate by InGaAsP Electroabsorption Modulator," *IEEE Electron. Lett.*, 28(1992), pp. 1007–1008.

[5] Smith, K., et al., "Totally Integrated Erbium Fibre Soliton Laser Pumped by Laser Diode, *IEEE Electron. Lett.*, 27(1991), pp. 244–246.

[6] Kodama, Y., and A. Hasegawa, "Amplification and Reshaping of Optical Solitons in Glass Fiber I–II," *Optics Lett.*, 7(1082), pp. 285–287, 339–341.

[7] Hasegawa, A., "Numerical Study of Optical Soliton Transmission Amplified Periodically by the Stimulated Raman Process," *Appl. Optics*, 23(1984), pp. 3302–3309.

[8] Zakharov, V. E., and A. B. Shabat, "Exact Theory of Two Dimensional Self Modulation of Waves in Nonlinear Media," *Sov. Phys. JETP*, 34(1972), pp. 118–126.

[9] Born, M., and E. Wolf, *Principles of Optics*, London: Pitman Press, 1964.

[10] Nakashima, T., et al., "Theoretical Limit of Repeater Spacing in an Optical Transmission Line Utilizing Raman Amplification," *IEEE/OSA J. Lightwave Techn.*, LT-4(1986), pp. 1267–1272.

[11] Kodama, Y., and A. Hasegawa, "Amplification and Reshaping of Optical Solitons in Glass Fiber— III. Amplifiers With Random Gain," *Optics Lett.*, 8(1083), pp. 342–344.

[12] Molemauer, L., et al., "Long Distance Soliton Propagation Using Lumped Amplifiers and Dispersion Shifted Fiber," *IEEE/OSA J. Lightwave Tech.*, LT-9(1991), pp. 194–197.

[13] Smith, K., and L. Molenauer, "Experimental Observation of Soliton Interaction Over Long Fiber Paths: Discovery of Long Range Interaction," *Optics Lett.*, 14(1989), pp. 1284–1286.

[14] Cvijetic, M., "Bit Error Rate Evaluation for Nonlinear Propagation in Optical Fibers," *Optical Quantum Electron.*, 22(1990), pp. 285–291.

[15] Gordon, J. P., and H. A. Haus, "Random Walk of Coherently Amplified Solitons in Optical Fiber Transmission," *Optics Lett.*, 11(1986), pp. 665–667.

[16] Chraplyvy, A. R., "Limitations on Lightwave Communications Imposed Optical-Fiber Nonlinearities," *IEEE/OSA J. Lightwave Techn.*, LT-8(1990), pp. 1548–1557.

[17] Mahlein, H. F., "Crosstalk Due to Stimulated Raman Scattering in Single-Mode Fibers for Optical Communication in Wavelength Division Multiplex Systems," *Optical Quantum Electron.*, 16(1984), pp. 409–425.

[18] Bracket, C. A., "Dense Wavelength Division Multiplexing Networks," *IEEE J. Select. Areas*, 8(1990), pp. 948–964.

[19] Kaminov, I. P., "FSK With Direct Selection in Optical Multiple Access FDM Networks," *IEEE J. Select. Areas*, 8(1990), pp. 1005–1013.

[20] Kuznetsov. M.. "Design of Widely Tunable Semiconductor Three Branch Lasers," *IEEE/OSA J. Lightwave Techn.*, LT-12(1994), pp 2100–2106.

[21] Shibata, N., et al., "Transmission Limitations Due to Fiber Nonlinearities in Optical FDM Systems," *IEEE J. Select. Areas*, 8(1990), pp. 1068–1077.

[22] Fishman, D. A., and J. A. Nagel, "Degradations Due to Stimulated Brillouin Scattering in Multigigabit Intensity Modulated Fiber Optic Systems," *IEEE/OSA J. Lightwave Techn.*, 11(1993), pp. 1721–1727.

[23] Inoue, K., "Fiber Four Wave Mixing Suppression Using Two Incoherent Polarized Lights," *IEEE/OSA J. Lightwave Techn.*, 11(1993), pp. 2116–2122.

[24] Inoue, K., "Suppression Technique for Fiber Four Wave Mixing Using Optical Multi/Demultiplexers and a Delay Line," *IEEE/OSA J. Lightwave Techn.*, 11(1993), pp. 455–461.

[25] Yoshizave, N., and T. Imai, "Stimulated Brillouin Scattering Suppression by Means of Applying Strain Distribution to Fiber With Cabling," *IEEE/OSA J. Lightwave Techn.*, 11(1993), pp. 1518–1522.

[26] Radgale, C. M., et al., "Narrowband Fiber Grating Filters," *IEEE J. Select. Areas*, 8(1990), pp. 1146–1150.

[27] Smith, D. A., et al. "Integrated Optic Acoustically Tunable Filters for WDM Networks," *IEEE J. Select. Areas*, 8(1990), pp. 1151–1159.

[28] Taga, H., et al., "Long Distance Multichannel WDM Transmission Experiments Using Er-Doped Fiber Amplifers," *IEEE/OSA J. Lightwave Techn.*, LT-12(1994), pp. 1448–1453.

[29] Nakazava, M., et al., "10 Gbit/s 1200 km Error Free Soliton Data Transmission Using Erbium Doped Fibre Amplifiers," *Electron. Lett.*, 28(1992), pp. 817–818.

[30] Mollenauer, L. F., et al. "Demonstration of Error Free Soliton Transmission Over More Than 15000 km at 5 Gbit/s, "Single Channel, and Over 11000 km at 10 Gbit/s in Two Channel WDM," *Electron. Lett.*, 28(1992), pp. 792–794.

Appendixes

APPENDIX A: SNR IN AN IDEAL OPTICAL RECEIVER

The average value of the current electrical signal at the output of an ideal photodetector is given as [1]

$$I_s = \frac{qP_s}{hf} \qquad (A.1)$$

where q is the electron charge, P_s is the average power of the incoming optical signal, h is Planck's constant, and f is the carrier frequency of optical signal. The signal power at load resistance R_L is given as

$$S = R_L I_s^2 = R_L \left(\frac{qP_s}{hf}\right)^2 \qquad (A.2)$$

The power of shot noise is (see Appendix F)

$$N = \overline{i^2} R_L = R_L (2qI_s B) \qquad (A.3)$$

where $\overline{i^2}$ is the mean-square value of the shot noise current. The mean-square value of the shot noise current in (A.3) is expressed as a product of the signal current, I_s, and the equivalent bandwidth, B, of the baseband filter (see Appendix B). Thus, the SNR in an ideal optical receiver becomes

$$SNR = S/N = \frac{I_s}{2qB} \qquad (A.4)$$

APPENDIX B: WHITE AND THERMAL NOISE

White noise, described by a function $n(t)$, possesses the constant power spectral density function $\Phi(\omega) = \nu_0/2$ [2]. When white noise passes through a linear system with transfer function $H(\omega)$, the power spectral density of the output noise becomes

$$\Phi_n(\omega) = \nu_0|H(\omega)|^2/2 \tag{B.1}$$

The total power of the output noise is

$$N = \int_{-\infty}^{\infty} \Phi_n(\omega)d\omega = \nu_0\int_0^{\infty}|H(\omega)|^2d\omega \tag{B.2}$$

If the linear system is an ideal low-pass filter with bandwidth B, the output noise power is given as

$$N = \nu_0|H(0)|^2B \tag{B.3}$$

where $H(0)$ presents the maximum value of the filter transfer function. The equivalent bandwidth B of noise is defined by (B.2) and (B.3), that is,

$$B = \frac{1}{|H(0)|^2}\int_0^{\infty}|H(\omega)|^2d\omega \tag{B.4}$$

The power spectral density of thermal noise generated by a resistor is defined as [3]

$$\Phi_T(\omega) = 2kTG\frac{\beta^2}{\beta^2 + \omega^2} \tag{B.5}$$

where k is Boltzmann's constant, T is the absolute temperature. G is the conductance of the resistor, and β is the average number of electron-lattice crashes. Since the parameter β is about 10^{14} for room temperature. the power spectral density of thermal noise is approximately constant and has the value $\Phi_T \simeq 2kTG$.

APPENDIX C: NARROWBAND NOISE

According to Appendix B, thermal noise can be approximated by white noise. Since thermal noise has the Gaussian distribution of the instantaneous values, it is commonly taken that white noise is the Gaussian process, as well [2]. Rice's discrete model of noise [4] assumed that the spectrum of white noise consists of an infinite number of pure spectral components, with equal amplitudes and uniform distribution of initial phases.

The individual spectral components of noise at the output of a narrowband filter will not have equal amplitudes. Since the component with the frequency equal to the filter central frequency, ω_c, will have the highest amplitude, it can be used as a referent. The frequencies of other components can be expressed as the sum of referent frequency steps and an integer number of differential frequency steps. Thus, the output noise becomes

$$n(t) = \sum_{k=-\infty}^{\infty} A_k \cos[(\omega_c + k\delta\omega)t + \varphi_k] \tag{C.1}$$

where A_k is the amplitude, and φ_k is the random phase of the kth component, while $\delta\omega$ represents the spectral distance between neighboring individual components.

The values A_k can be found from the condition of equal powers in time and frequency domains, that is,

$$\overline{n(t)^2} = \int_0^\infty \Phi_n(f)df = \sum_{k=-\infty}^{\infty} \frac{1}{2}A_k^2 \tag{C.2}$$

Since the frequency step $\delta\omega$ is very small, the integral in (C.2) can be replaced by the sum

$$\overline{n(t)^2} = \sum_{k=-\infty}^{\infty} \Phi_n(f_c + k\delta f)\delta f \tag{C.3}$$

Thus, the coefficient A_k corresponding to the amplitude of kth discrete noise component, has the value

$$A_k = \sqrt{2\Phi_n(f_c + k\delta f)\delta f} \tag{C.4}$$

Hence, Rice's expression for narrowband noise takes the form

$$n(t) = x(t)\cos \omega_c t - y(t)\sin \omega_c t \tag{C.5}$$

where

$$x(t) = \sum_{k=-\infty}^{\infty} \sqrt{2\Phi_n(f_c + k\delta f)\delta f}\cos(k\delta\omega t + \varphi_k) \tag{C.6}$$

and

$$y(t) = \sum_{k=-\infty}^{\infty} \sqrt{2\Phi_n(f_c + k\delta f)\delta f} \, \sin(k\delta\omega t + \varphi_k) \qquad (C.7)$$

The even and odd components of the noise are the Gaussian processes with mean values equal to zero and with average powers equal to

$$\overline{x(t)^2} = \overline{y(t)^2} = \sum_{k=-\infty}^{\infty} \Phi_n(f_c + k\delta f)\delta f \qquad (C.8)$$

From (C.3) it follows that

$$\overline{x(t)^2} = \overline{y(t)^2} = \overline{n(t)^2} = \sigma^2 \qquad (C.9)$$

The slowly varying processes $x(t)$ and $y(t)$ are mutually orthogonal, so they are noncorrelated. Hence, the probability density functions of processes $x(t)$ and $y(t)$ are the Gaussian functions

$$w(x) = \frac{1}{\sqrt{2\pi}\sigma} e^{-x^2/2\sigma^2} \qquad (C.10)$$

and

$$w(y) = \frac{1}{\sqrt{2\pi}\sigma} e^{-y^2/2\sigma^2} \qquad (C.11)$$

The envelope of the narrowband noise is defined as

$$r(t) = \sqrt{x(t)^2 + y(t)^2} \qquad (C.12)$$

while its phase has the form

$$\varphi(t) = \text{arctg}\frac{y(t)}{x(t)} \pm m\pi, \quad m = 0, 1, 2, \ldots \qquad (C.13)$$

Thus, the narrowband noise can be expressed in the form

$$n(t) = r(t)\cos[\omega_c t + \varphi(t)] \qquad (C.14)$$

To find the distribution functions for amplitude and phase of the narrowband noise, the joined probability density function $w(x, y)$ should be transformed into the joined probability density function $w_n(r, \varphi)$ by the equality $w(x, y)dxdy = w_n(r, \varphi)drd\varphi$.

Since the processes $x(t)$ and $y(t)$ are statistically independent, the joined probability can be expressed as the product $w(x)w(y)$. Further, the differential area $dxdy$ is equal to the differential area $rdrd\varphi$. Then we have that

$$\frac{e^{-r^2/2\sigma^2}}{2\pi\sigma^2} rdrd\varphi = w_n(r,\ \varphi)drd\varphi \qquad (C.15)$$

The joined probability distribution function of instantaneous amplitude and instantaneous phase of narrowband noise is

$$w_n(r,\ \varphi) = \frac{1}{2\pi}\frac{r}{\sigma^2}e^{-r/2\sigma^2} \qquad (C.16)$$

The probability density functions of amplitude and phase can be easily found from (C.16) by corresponding integration. Thus, the probability density function of instantaneous amplitude is found by integration of (C.16) over the entire phase region, and we have that

$$w(r) = \int_0^{2\pi} w_n(r,\ \varphi)d\varphi = \frac{r}{\sigma^2}\exp\left(-\frac{r}{\sigma^2}\right) \qquad (C.17)$$

Equation (C.17) presents the well-known Rayleigh distribution function, shown in Figure C.1. This function can be used for the evaluation of the mean value and the mean-square value of the narrowband noise envelope. The envelope mean value is

$$\bar{r} = \int_0^\infty rw(r)dr = \sigma\sqrt{\pi/2} = 1.25\sigma \qquad (C.18)$$

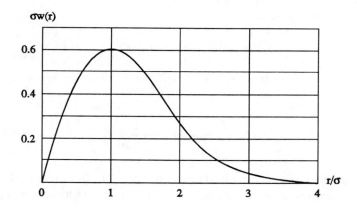

Figure C.1 Rayleigh distribution function of narrowband noise envelopes.

while the mean-square value becomes equal to

$$\overline{r^2} = \int_0^\infty r^2 w(r)dr = 2\sigma^2 \tag{C.19}$$

The variance of the instantaneous values of the noise envelope can be expressed as

$$\sigma_r^2 = \overline{r^2} - \overline{r}^2 = 0.43\sigma^2 \tag{C.20}$$

The probability density function of the narrowband noise phase can be found by the integration of (C.16) from zero to infinity, that is,

$$w(\varphi) = \int_0^\infty w_n(r, \varphi)dr = \frac{1}{2\pi}, \quad 0 \le \varphi \le 2\pi \tag{C.21}$$

APPENDIX D: NARROWBAND SIGNAL AND NARROWBAND NOISE

If the sum of a signal and noise passes through the narrowband filter (with central frequency f_c and bandwidth B), it is necessary to know the characteristics of the output process. It can be assumed that the output filtrated process has the form

$$f(t) = A \cos \omega_c t + n(t) \tag{D.1}$$

where the signal component is expressed by a simple cosine function. According to (C.5), (D.1) can be rewritten as

$$f(t) = [x(t) + A]\cos \omega_c t - y(t)\sin \omega_c t \tag{D.2}$$

The even component $z(t) = [x(t) + A]$ will be the Gaussian random variable, with the probability density function

$$w(z) = w(z - A) = \frac{1}{\sqrt{2\pi\sigma^2}}e^{-(z-A)^2/2\sigma^2} \tag{D.3}$$

The additive process of mixed signal and noise at the output of the narrowband filter can be expressed in a form equivalent to (C.14), with the envelope

$$r(t) = \sqrt{z(t)^2 + y(t)^2} \tag{D.4}$$

and phase

$$\varphi(t) = \text{arctg}\frac{y(t)}{z(t)} \pm m\pi, \quad m = 0, 1, 2, \ldots \tag{D.5}$$

According to the rule for transformation of joined probability distribution functions [5], the following equation can be written:

$$w_n(r, \varphi)drd\varphi = w(z)w(y)dzdy = \frac{1}{2\pi\sigma^2}e^{-[(z-A)^2+y^2]/2\sigma^2} \tag{D.6}$$

Since

$$z = r \cos \varphi \text{ and } y = r \sin \varphi \tag{D.7}$$

the joined distribution function takes the form

$$w_n(r, \varphi) = \frac{1}{2\pi\sigma^2}e^{-\frac{r^2 + A^2}{2\sigma^2}}e^{\frac{rA \cos \varphi}{\sigma^2}} \tag{D.8}$$

The probability density function of the instantaneous envelope of the output process, is found by the integration

$$w(r) = \int_0^{2\pi} w_n(r, \varphi)d\varphi = \frac{r}{\sigma^2}e^{\frac{r^2 + A^2}{2\sigma^2}}\left(\frac{1}{2\pi}\int_0^{2\pi} e^{\frac{rA \cos \varphi}{\sigma^2}} d\varphi\right) \tag{D.9}$$

The term in parentheses is, in fact, the modified Bessel function of first kind and zero-th order $I_0(rA)$ [5], so we have

$$w(r) = \frac{r}{\sigma^2} \exp\left(-\frac{r^2 + A^2}{2\sigma^2}\right)I_0\left(\frac{rA}{\sigma^2}\right) \tag{D.10}$$

Equation (D.10) presents the well-known Rice's probability distribution function, which is shown in Figure D.1. When amplitude A tends to zero value, we have that $I_0(0) = 1$, and the Rice's function takes the form of a Rayleigh function. When the ratio A/σ increases, the Rice's-function shape approaches the Gaussian shape. Namely, for the large values of an argument (i.e., for $rA/\sigma^2 \gg 1$), we have

$$I_0(u) \simeq \frac{1}{\sqrt{2\pi\sigma^2}} \exp(u) \tag{D.11}$$

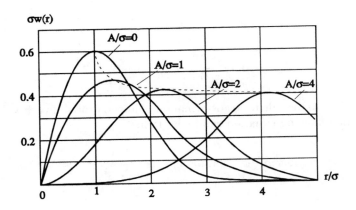

Figure D.1 Rice's distribution function of narrowband signal-noise process envelopes.

and the Rice's distribution becomes

$$w(r) = \frac{r}{\sigma} \frac{1}{\sqrt{2\pi rA}} \exp\left[-\frac{(r-A)^2}{2\sigma^2}\right] \tag{D.12}$$

The nonexponential part of (D.12) has considerable influence on the function $w(r)$ only in the vicinity of point $r = A$, so the approximation $r \simeq A$ can be used for the nonexponential part. Then the Rice's function takes the classical Gaussian form

$$w(r) = \frac{1}{\sqrt{2\pi\sigma^2}} \exp\left[-\frac{(r-A)^2}{2\sigma^2}\right] \tag{D.13}$$

To obtain the probability density function for the instantaneous phase of the narrowband signal-noise process, the (D.8) must be integrated from zero to infinity. In such a way, the following equation is obtained:

$$w(\varphi) = \int_0^\infty w_n(r, \varphi)dr = \frac{1}{2\pi\sigma^2}\exp\left(-\frac{A}{2\sigma^2}\right)I_1 \tag{D.14}$$

Parameter I_1 denotes the definite integral

$$I_1 = \int_0^\infty r \exp\left(-\frac{r^2 - 2Ar\cos\varphi}{2\sigma^2}\right)dr \tag{D.15}$$

By solving (D.15) [2], the expression for function $w(\varphi)$ takes the form

$$w(\varphi) = \frac{1}{2\pi} \exp(-\rho^2) + \frac{1}{2\sqrt{\pi}} \rho \cos \varphi \exp(-\rho^2 \sin^2 \varphi)\left[1 + \Phi(\rho \cos \varphi) \right] \quad \text{(D.16)}$$

where the parameter ρ defines the signal-to-noise power ratio, that is,

$$\rho^2 = \frac{A^2}{2\sigma^2} \tag{D.17}$$

while function $\Phi(z)$ is the Gaussian error function, which has the form [5]

$$\Phi(z) = \frac{2}{\sqrt{\pi}} \int_0^z \exp(-u^2)du \tag{D.18}$$

In the absence of a signal, (D.16) takes the constant value in the interval from 0 to 2π. For the large values of SNR ($\rho^2 \gg 1$), the phase of the additive process at the output of the narrowband filter will be most probably in the vicinity of the point $\varphi = 0$. The function $w(\varphi)$ takes the Gaussian form in the vicinity of point $\varphi = 0$. In such a case, the approximations $\sin \varphi \simeq \varphi$ and $\cos \varphi \simeq 1$ can be applied, so we have

$$\Phi(\rho \cos \varphi) \simeq \Phi(\rho) \simeq 1, \quad \rho^2 \gg 1 \tag{D.19}$$

and

$$w(\varphi) \simeq \frac{1}{\sqrt{\pi/\rho^2}} \exp(- \rho^2\varphi^2) \tag{D.20}$$

APPENDIX E: ERROR PROBABILITY EVALUATION IN IM/DD SYSTEMS

Lightwave systems with intensity modulation and direct detection are characterized by transmission of unipolar pulses in the frequency baseband, in the presence of additive noise. The additive noise is generated mainly in the optical receiver. A transmitted binary pulse train has the form

$$s_p(t) = \sum_n a_n h_p(t - nT_d), \quad a_n = 0 \text{ or } 1 \tag{E.1}$$

where $h_p(t)$ presents an individual pulse of rectangular shape and with width equal to the digit interval length T_d. This pulse propagates through an optical-fiber line, changing its own shape. The pulse shape $h(t)$ after the receiver low-pass filter is very close to the Gaussian curve with maximal value S at the middle of the digit interval [1]. Thus, the time moment $t = T_d/2$ can be taken as a referent for sampling and decision. An additive noise $n(t)$ is added to the received signal; hence, the received waveform can be expressed as the sum

$$v(t) = s(t) + n(t) \tag{E.2}$$

When the state "1" was sent the corresponding output voltage at the sampling time is $v_1(t_0)$, while the voltage $v_0(t_0)$ corresponds to the sent "0" state. The registered voltage levels are given as

$$v_1(t_0) = S + n(t_0) \tag{E.3a}$$

$$v_0(t_0) = n(t_0) \tag{E.3b}$$

It is assumed that if the noise $n(t)$ is a Gaussian process [6], then the assemblage of samples, $v_1(t_0)$, will also be a Gaussian process with some mean value S. The probability density function for such a case is

$$w_1(v) = \frac{1}{\sqrt{2\pi}\sigma_1} \exp\left(- \frac{(v - S)^2}{2\sigma_1^2}\right) \tag{E.4}$$

where σ_1^2 is the variance (or average power per assemblage) of the noise at the output of the receiver filter when "1" was sent. At the same time, the probability density function for elements $v_0(t_0)$ has the form

$$w_0(v) = \frac{1}{\sqrt{2\pi}\sigma_0} \exp\left(- \frac{v^2}{2\sigma_0^2}\right) \tag{E.5}$$

The error probability in a binary IM/DD system is defined as

$$P_{e,2} = P(0)P(1/0) + P(1)P(0/1) \tag{E.6}$$

The *a priori* probabilities that either "1" or "0" will be sent are nearly equal for line-coded signals, so we have that $P(0) = P(1) = 0.5$. The conditional probability that the output voltage will be detected as a "1" state if a "0" state was sent is

$$P(1/0) = \int_b^\infty w_0(v)dv = 0.5\left[1 - \Phi\left(\frac{b}{\sqrt{2}\sigma_0}\right)\right] \tag{E.7}$$

where $\Phi(z)$ presents the Gaussian error function [see (D.18)], and b is the decision threshold.

The conditional probability that the output voltage is less than threshold b when "1" was sent is

$$P(0/1) = \int_{-\infty}^{b} w_1(v)dv = 0.5\left[1 + \Phi\left(\frac{b-S}{\sqrt{2}\sigma_1}\right)\right] \tag{E.8}$$

Thus, the expression for error probability in a binary IM/DD lightwave system becomes

$$P_{e,2} = 0.5\left[1 - 0.5\Phi\left(\frac{S-b}{\sqrt{2}\sigma_1}\right) - 0.5\Phi\left(\frac{b}{\sqrt{2}\sigma_0}\right)\right] \tag{E.9}$$

The optimum value of the threshold, b_{opt}, can be found from

$$\frac{S-b}{\sigma_1} = \frac{b}{\sigma_2} \tag{E.10}$$

So we have

$$b_{opt} = \frac{\sigma_0 S}{\sigma_0 + \sigma_1} \tag{E.11}$$

The final expression for error probability in an optimum detection scheme has the form

$$P_{e,2} = 0.5\left[1 - \Phi\left(\frac{1}{\sqrt{2}}\frac{S}{\sigma_1 + \sigma_0}\right)\right] \tag{E.12}$$

or the more familiar form

$$P_{e,2} = 0.5 \; \text{erfc}\left(\frac{1}{\sqrt{2}}\frac{S}{\sigma_1 + \sigma_0}\right) \tag{E.13}$$

APPENDIX F: PIN AND AVALANCHE PHOTODIODES NOISE

The photocurrent generated in a PIN photodiode for incoming optical power, $p(t)$, is

$$i(t) = \frac{\eta q}{hf}p(t) \tag{F.1}$$

where η is the quantum efficiency of the photodiode, q is the electron charge, h is Planck's constant, and f is the carrier frequency of the optical signal. The total power of the current signal is a time-averaged value of the current square, or

$$S = \overline{i^2(t)} \tag{F.2}$$

If it is assumed that $p(t)$ is a sine-modulated signal,

$$p(t) = P_s(1 + \sin \omega t) \tag{F.3}$$

the signal power becomes

$$S = \frac{1}{2}\left[P_s\frac{\eta q}{hf}\right]^2 = \frac{I_s^2}{2} \tag{F.4}$$

The parameter I_s represents the mean value of the photocurrent signal.

Quantum, or shot, noise generated in a photodiode can be considered white noise. The total noise power can be expressed as the product of the power spectral density of the noise and the equivalent bandwidth of the baseband filter (see Appendix B):

$$N_s = \overline{[i(t) - <i(t)>]^2} = 2qI_sB \tag{F.5}$$

The parameter $<i(t)>$ presents an averaged current value over the assemblage of generated electron-hole pairs.

Another kind of noise in a PIN photodiode is the dark current due to thermal effect. There are two kinds of dark current: surface dark current, i_{ds}, and bulk dark current, i_{db}. According to (F.5), dark current noise powers can be expressed as

$$N_{ds} = 2qi_{ds}B \tag{F.6}$$

and

$$N_{db} = 2qi_{db}B \tag{F.7}$$

The surface current component can be suppressed by proper structure of the photodiode, so the value of (F.6) is commonly neglected.

An avalanche photodiode (APD) is characterized by the avalanche amplification of the primarily generated electron-hole pairs by the impact ionization process. The amplification can be characterized by the multiplication coefficient, $M = I_M/I_s$, where I_M presents the output current from the APD, while I_s is the primary unmultiplied photocurrent defined in (F.1). The power of the current signal generated in an APD can be expressed as

$$S = \overline{[i(t)\overline{M}]^2} = \overline{i^2(t)}\overline{M}^2 = \frac{1}{2}I_s^2\overline{M}^2 \tag{F.8}$$

The total power of quantum noise, consisting of shot noise and avalanche multiplication noise, can be determined as

$$N_s = \overline{[i(t)M - <i(t)M>]^2} \tag{F.9}$$

The averaged value $<M>$ over the assemblage of primary generated electron-hole pairs will be, in general, different from value $\overline{M}$. Because of that, the following relation is valid:

$$<M^2> = F(<M>)<M>^2 = F(\overline{M})\overline{M}^2 \tag{F.10}$$

where $F(\overline{M})$ is some increasing function of argument $\overline{M}$, which is higher than or equal to unity. Now according to (F.5) and (F.10), the power of APD quantum noise takes the form

$$N_s = 2q\left[P_sF(\overline{M})\overline{M}^2\frac{\eta q}{hf}\right]B = 2qI_sBF(\overline{M})\overline{M}^2 \tag{F.11}$$

The bulk dark current in an APD causes noise with the power

$$N_{db} = 2qi_{db}BF(\overline{M})\overline{M}^2 \tag{F.12}$$

The APD noise power caused by surface dark current is given by (F.6), because there is no avalanche multiplication.

The function $F(\overline{M})$, commonly called the *excess noise factor*, is most often represented by the function

$$F(\overline{M}) = M^x \tag{F.13}$$

where the coefficient x depends on the kind of material. Its value is 0.3 to 0.4 for a silicium APD, 0.9 to 1.0 for a germanium APD, and 0.7 to 0.8 for an APD based on an InGaAs-AsP structure.

The more exact model of function $F(\overline{M})$ takes into account the different behavior of electrons and holes during the ionization process in a strong electrical field. When the electron influence in such a process is dominant, the following expression is valid:

$$F_n(\overline{M}) = \overline{M}\left[1 - \frac{(1-k)(\overline{M}-1)^2}{\overline{M}^2}\right], \; k \ll 1 \tag{F.14}$$

while for the hole-dominant effect, it is

$$F_p(\overline{M}) = \overline{M}\left[1 - \frac{(k-1)(\overline{M}^2-1)^2}{\overline{M}^2 k}\right], \; k \gg 1 \tag{F.15}$$

where k is the so-called *ionization factor ratio*. Its value is 0.015 to 0.03 for a silicium APD, 0.7 to 0.9 for a germanium APD, and 0.3 to 0.5 for an APD based on an InGaAs-AsP structure.

APPENDIX G: NOISE IN IM/DD OPTICAL RECEIVERS

To find the total noise in an IM/DD optical receiver, we can use Figure 4.1. The incoming pulse train of the optical signal, $p(t)$, can be expressed as

$$p(t) = \sum_{n=-\infty}^{\infty} b_n h_f(t - nT_d) \tag{G.1}$$

where b_n takes the value 1 or 0, while $h_f(t)$ represents the time shape of individual optical pulses. The current response for the input optical signal $p(t)$ is

$$i(t) = \Re\overline{M}\sum_{n=-\infty}^{\infty} b_n h_f(t - nT_d) \tag{G.2}$$

where $\Re$ is the photodiode responsivity, and $\overline{M}$ is the coefficient of avalanche amplification ($\overline{M} = 1$ for a PIN photodiode). The averaged voltage at the output of the receiver equalizer is the sum of individual pulses, $h(t)$, which are obtained from the pulses $h_f(t)$ due to the influence of transfer functions of the receiver amplifier and the receiver equalizer. So we have

$$v(t) = \sum_{n=-\infty}^{\infty} b_n h(t - nT_d) \tag{G.3}$$

The shape of an individual output pulse is given by

$$h(t) = \Re\overline{M}Ah_f(t) * h_A(t) * h_E(t) \tag{G.4}$$

where A and $h_A(t)$ are the amplification coefficient and the impulse response of the receiver amplifier, respectively, while $h_E(t)$ represents the impulse response of the equalizer.

The expression for noise voltage at the equalizer output, according to Figure 4.1, is [6]

$$\overline{v_n^2(t)} = 2q\overline{i}\ \overline{M}^{2+x}BR^2A^2 + \frac{4k\Theta}{R_L}BR^2A^2 + 2G_IBR^2A^2 + 2G_EB_EA^2 \tag{G.5}$$

The current $\overline{i}$ corresponds to an averaged value of the signal current within the bit interval T_d. The bandwidth of the equivalent filter for the voltage noise source differs from the bandwidth of the equivalent filter for current noise source, because of their different connection in equivalent scheme. The bandwidth of the equivalent filter for voltage source noise is determined by

$$2B_E = \frac{1}{|H_E(0)|^2}\int_0^\infty |H_E(f)|^2 df = \frac{R^2|H_f(0)|^2}{|H(0)|^2}\int_0^\infty \left|\frac{H(f)}{H_f(f)}\left(\frac{1}{R} + j2\pi fC\right)\right|^2 df \tag{G.6}$$

where cardinal H-letters concern corresponding Fourier transforms of the impulse responses, while R and C present the equivalent resistance and equivalent capacitance of the scheme from Figure 4.1, respectively. Equation (G.6) can be transformed into a more suitable form by introducing the dimensionless parameters $\tau = t/T_d$ and $\varphi = fT_d$, so we have [7]

$$\overline{v_n^2(t)} = \frac{R^2A^2}{T_d}\left[q\overline{i}\ \overline{M}^{2+x} + \frac{2k\Theta}{R_L} + G_I + \frac{G_E}{R^2}\right]I_2 + \frac{(2\pi RC)^2A^2}{T_d^3}G_EI_3 \tag{G.7}$$

where I_2 and I_3 are the so-called *Personic's integrals*, which have the form

$$I_2 = \int_{-\infty}^\infty \left|\frac{H'(\varphi)}{H_f'(\varphi)}\right|^2 d\varphi \tag{G.8}$$

and

$$I_3 = \int_{-\infty}^\infty \left|\frac{H'(\varphi)}{H_f'(\varphi)}\right|^2 \varphi^2 d\varphi \tag{G.9}$$

The functions $H'(\varphi)$ and $H_f'(\varphi)$ present the Fourier transforms of impulse responses $h(\tau)$ and $h_f(\tau)$, respectively. The values of integrals corresponding to the real situations are about unity for I_2 and about 0.3 for I_3.

APPENDIX H: OPTIMAL DIGITAL RECEIVERS

The main task of a binary digital receiver is the proper choice of one of two possible hypotheses:

$$H_0: x(t) = s_0(t) + n(t), \; < t < T_d \tag{H.1}$$

or

$$H_1: x(t) = s_1(t) + n(t), \; 0 < t < T_d \tag{H.2}$$

where $s_0(t)$ and $s_1(t)$ are the binary elementary signals, which can appear in one-digit interval (from 0 to T_d), while $n(t)$ represents the additive Gaussian noise with zero mean value and variance σ^2. The power spectral density of filtrated Gaussian noise is ν_0. The optimum digital receiver chooses one of two possible hypotheses with minimal risk. The receiver should form the likelihood of ratio Λ on the basis of sampled values of process $x(t)$ and then compare it with some threshold value Λ_0 [1]. To make the true decision, it must be assumed that the process $x(t)$ is sampled n times within the considered time interval after the receiver filter. The false hypothesis, H_0, at the moment $t = t_k$, where $1 \le k \le n$, gives the next value of the received process:

$$x_k(t_k) = x_k = s_{0k} + n_k \tag{H.3}$$

while the true hypothesis, H_1, gives the value

$$x_k(t_k) = x_k = s_{1k} + n_k \tag{H.4}$$

The likelihood ratio is defined as

$$\Lambda(X) = \frac{w_1(X)}{w_0(X)} = \frac{w_1(x_1, x_2, x_3, \ldots, x_n)}{w_0(x_1, x_2, x_3, \ldots, x_n)} \tag{H.5}$$

where $w_1(X)$ and $w_0(X)$ are the functions of joined probability densities of vector X. The hypothesis H_0 will be chosen in the decision stage of the receiver if $\Lambda(X) < \Lambda_0$, while the hypothesis H_1 will be chosen if $\Lambda(X) > \Lambda_0$. The choice of hypothesis means, in fact, the satisfaction of the condition

$$G = \int_0^{T_d} x(t)s_1(t)dt - \int_0^{T_d} x(t)s_0(t)dt + \frac{1}{2}\int_0^{T_d} [s_0^2(t) - s_1^2(t)]dt \ge 0 \tag{H.6}$$

If the probability density functions of process G, under the hypotheses H_0 and H_1, have the values $w_0(G)$ and $w_1(G)$, respectively, the minimal error probability is defined as

$$P_e = P(H_0)P(D_1/H_0) + P(H_1)P(D_0/H_1)$$
$$= \frac{1}{2}\int_0^\infty w_0(G)dG + \frac{1}{2}\int_{-\infty}^0 w_1(G)dG \tag{H.7}$$

where D_0 and D_1 denote the decisions corresponding to the first and the second hypothesis, respectively.

Since G is the normal Gaussian variable, it is necessary to find only its mean value and variance. The mean value for hypothesis H_1 is

$$\overline{G}/H_1 = \frac{1}{2}\int_0^{T_d} [s_0(t) - s_1(t)]^2 dt \tag{H.8}$$

while the mean value for hypothesis H_0 is

$$\overline{G}/H_0 = -\frac{1}{2}\int_0^{T_d} [s_0(t) - s_1(t)]^2 dt \tag{H.9}$$

The variance of the process is the same for both hypotheses and has the value

$$\sigma_G^2 = \nu_0 \int_0^{T_d} [s_0(t) - s_1(t)]^2 dt \tag{H.10}$$

By introducing the parameters $\overline{E}$ and κ, which denotes the mean energy of both states and the cross-correlational coefficient, respectively, (H.8) and (H.9) can be written in a more suitable form. The parameters $\overline{E}$ and κ are defined as

$$\overline{E} = \frac{1}{2}\int_0^{T_d} [s_0^2(t) + s_1^2(t)]dt \tag{H.11}$$

and

$$\kappa = \frac{\int_0^{T_d} [s_0(t)s_1(t)]dt}{\overline{E}} \tag{H.12}$$

Thus, (H.8) and (H.9) take the forms

$$\overline{G}/H_1 = \overline{E}(1 - \kappa) \tag{H.13}$$

and

$$\overline{G}/H_0 = -\overline{E}(1 - \kappa) \tag{H.14}$$

while the variance of the process G becomes

$$\sigma_G^2 = 2\nu_0 \overline{E}(1 - \kappa) \tag{H.15}$$

Finally, the equation for error probability in an optimal digital receiver takes the form

$$P_e = \text{erfc} \sqrt{\frac{\overline{E}(1 - \kappa)}{4\nu_0}} \tag{H.16}$$

Equation (H.16) is used in Chapter 2 for the evaluation of minimal error probabilities in coherent optical receivers.

APPENDIX I: NUMBER OF LASER EIGENMODES

The number of the states of freedom of an electromagnetic field, which correspond to the number of modes, can be evaluated in an appropriate closed space. It is most convenient to choose a cube as that space and evaluate the possible number of eigenoscillations of the field in the cube. The eigenoscillations form the stationary waves in the direction of every edge in the cube (actually, in every axis direction), with the nodes on the cube sides. The frequency of these oscillations is determined from the condition valid for wave vector k [8], that is,

$$k^2 = (\omega/c)^2 = k_x^2 + k_y^2 + k_z^2 \tag{I.1}$$

where ω is the carrier frequency of the optical wave, and c is the light velocity in free space. The components of the wave vector in the cube space must satisfy the following conditions:

$$k_x = \frac{2\pi l}{L}; \; k_y = \frac{2\pi n}{L}; \; k_z = \frac{2\pi m}{L}; \; l, \, n, \, m = 1, \, 2, \, 3, \, \ldots \tag{I.2}$$

where L is the length of the cube edge. Since the distance between the neighboring order oscillations is $k_x = \Delta k_y = \Delta k_z = 2\pi/L$, each couple of eigenoscillations (with mutually orthogonal polarizations) occupies the space

$$\Delta V_k = \Delta k_x \Delta k_y \Delta k_z = (2\pi/L)^3 \qquad (I.3)$$

The number of eigenoscillations in a ball with radius k and with the cubic $V_k = 2\pi k^3/3$ are

$$N_k = 2V_k/\Delta V_k = k^3 L^3/(3\pi^2) \qquad (I.4)$$

There are $n_k = k^3/(3\pi^2)$ eigenoscillations in the frequency region from 0 to $f = kc/(2\pi)$, in the cube with the cubic $V = L^3$. The spectral density, n_f, of eigenoscillations can be determined from the space density, n_k, of the eigenoscillations, that is,

$$n_f = \frac{dn_k}{df} = 8\pi f^2/c^3 \qquad (I.5)$$

Each eigenoscillation per one state of freedom has two energy reservoirs, the electrical field and the magnetic field. Under the uniform energy distribution over eigenoscillations, the energy per one eigenoscillation is equal to $W = k_B\Theta$, and the spectral density of energy is given as

$$\sigma = n_f k_B \Theta = 8\pi f^2 k_B \Theta/c^3 \qquad (I.6)$$

where k_B is Boltzmann's constant, and Θ is the absolute temperature.

APPENDIX J: GROUP DELAY AND CHROMATIC DISPERSION

The spectral components in the wavelength band over which the light source emits signals propagate along the optical fiber with different group velocities, v_g. Hence, they undergo the different group delays, τ_g, per unit length. It is valid that [1]

$$\frac{\tau_g}{L} = \frac{1}{v_g} = \frac{1}{c}\frac{d\beta}{dk} = -\frac{\lambda^2}{2\pi c}\frac{d\beta}{dk} \qquad (J.1)$$

where L is the transmission distance, c is the light velocity in free space, λ is the wavelength, β is the propagation constant, and k is the wave number. As the result of the difference in time delays, the optical signal pulse will spread out during its propagation along the optical fiber. For a narrow spectral linewidth of the source, the delay difference per unit wavelength is approximately $d\tau_g/d\lambda$, so the finite spectral linewidth, $\Delta\lambda$, causes the total delay difference

$$\tau = \frac{\Delta\lambda\, d\tau_g}{d\lambda} \qquad (J.2)$$

Now, according to (J.1) and (J.2), we have

$$\tau = -\frac{L\,\Delta\lambda}{2\pi c}\left(2\lambda\frac{d\beta}{d\lambda} + \lambda^2\frac{d^2\beta}{d\lambda^2}\right) \qquad (J.3)$$

The term $D = \tau/(L\Delta\lambda)$ is designated as the dispersion. The dispersion, D, can be expressed as the sum $D = D_m + D_w$, where

$$D_m = -\frac{\lambda^2}{2\pi c}\frac{d^2\beta}{d\lambda^2} \qquad (J.4)$$

presents material dispersion, while the term

$$D_w = -\frac{\lambda}{\pi c}\frac{d\beta}{d\lambda} \qquad (J.5)$$

presents waveguide dispersion. The evaluation of material and wavelength dispersion components can be made for an arbitrary carrier wavelength λ, by knowing that $k = 2\pi/\lambda$ and $\beta = 2\pi n(\lambda)/\lambda$.

REFERENCES

[1] Keiser, G., "Optical Fiber Communications," McGraw Hill, Tokyo, 1983.
[2] Lee, Y. W., "Statistical Theory of Communications," John Wiley and Sons, New York, 1960.
[3] Johnson, J. B., "Thermal agitation of electricity in conductors," *Phys. Rev.*, 32(1928), 97–109.
[4] Rice, S. O., "Mathematical analysis of random noise," *Bell Syst. Techn. J.*, 24(1945), 46–156.
[5] Korn, G., T. Korn, "Mathematical Handbook for Scientist and Engineers," McGraw Hill, London, 1961.
[6] Smith, D., I. Garret, "A simplified approach to digital optical receiver design," *Opt. Quant. Electron.*, 10(1978), 211–221.
[7] Personick, D., "Receiver design for digital fiber optic communication system," *Bell Syst. Techn. J.*, 52(1973), 843–886.
[8] Born, M., E. Wolf, "Principles of Optics," Pergamon Press, Oxford, 1964.

List of Acronyms

APD	avalanche photodiode
ASK	amplitude-shift keying
ATM	asynchronous transfer mode
BER	bit-error rate
CPFSK	continuous phase FSK
CPSK	coherent PSK
DBR	distributed Bragg reflector
DCPSK	differentially coherent PSK
DD	direct detection
DFB	distributed feedback
DPSK	differential PSK
FDM	frequency division multiplex
FET	field effect transistor
FSK	frequency shift keying
IF	intermediate frequency
IM	intensity modulation
MQW	multiple quantum well
MSK	minimum shift keying
OPLL	optical PLL
PIN	layers in photodiode (positive-intrinsic-negative)
PLL	phase-locked loop
PSK	phase-shift keying
RC	resistance capacity

SDH	synchronous digital hierarchy
SNR	signal-to-noise ratio
TE	transversal electric
TEM	transversal electromagnetic
VCO	voltage-controlled oscillator
WDM	wavelength division multiplex

About the Author

Milorad Cvijetic received his Ph.D. degree in electrical engineering from Belgrade University in 1984. He currently serves as a member of the scientific staff of Bell-Northern Research in Ottawa in the Advanced Technology Laboratory.

Dr. Cvijetic has authored more than 30 technical papers in the fields of digital transmission and optical communications, along with the text book *Digital Optical Communications*. He has taken part in numerous telecommunications conferences, some as a session chairman. He is a member of the New York Academy of Science and a member of IEEE.

Index

The Artech House Optoelectronics Library

For further information on these and other Artech House titles, contact:

Artech House
685 Canton Street
Norwood, MA 02062
617-769-9750
Fax: 617-769-6334
Telex: 951-659
email: artech@world.std.com

Artech House
Portland House, Stag Place
London SW1E 5XA England
+44 (0) 171-973-8077
Fax: +44 (0) 171-630-0166
Telex: 951-659
email: bookco@artech.demon.co.uk